Multiple Valued Logic:
Concepts and Representations

Multiple Valued Logic: Concepts and Representations
D. Michael Miller and Mitchell A. Thornton

ISBN: 978-3-031-79778-1   paperback
ISBN: 978-3-031-79779-8   ebook

DOI: 10.1007/978-3-031-79779-8

A Publication in the Springer series

SYNTHESIS LECTURES ON DIGITAL CIRCUITS AND SYSTEMS #12

Lecture #12
Series Editor: Mitchell A. Thornton, Southern Methodist University

Series ISSN

ISSN 1932-3166   print
ISSN 1932-3174   electronic

Multiple Valued Logic: Concepts and Representations

D. Michael Miller
University of Victoria, Canada

Mitchell A. Thornton
Southern Methodist University, USA

SYNTHESIS LECTURES ON DIGITAL CIRCUITS AND SYSTEMS #12

ABSTRACT

Multiple-Valued Logic: Concepts and Representations begins with a survey of the use of multiple-valued logic in several modern application areas including electronic design automation algorithms and circuit design. The mathematical basis and *concepts* of various algebras and systems of multiple-valued logic are provided including comparisons among various systems and examples of their application. The book also provides an examination of alternative *representations* of multiple-valued logic suitable for implementation as data structures in automated computer applications. Decision diagram structures for multiple-valued applications are described in detail with particular emphasis on the recently developed quantum multiple-valued decision diagram.

KEYWORDS:

Multiple-valued Logic, Quantum Logic, Reversible Logic, Decision Diagram, QMDD, Algebra

Contents

List of Figures

List of Tables

Preface

Multiple-Valued Logic Concepts and Representations provides an easily accessible overview of basic concepts and representations in multiple-valued logic (MVL). A survey of MVL ideas in terms of the underlying mathematics as well as applications in computer system design and implementation that allows for a concise introduction to the topic is included. Important concepts from early theoretical formulations to modern applications are discussed that should be of interest to readers who have some familiarity with discrete mathematics and the principles of digital system architecture.

The book begins with a survey of the use of MVL in several modern application areas including electronic design automation algorithms and circuit design. This portion of the book provides the motivation for the use of MVL through the use of example applications in current use. These applications include algorithms, logic circuits based on conventional devices, and logic system architectures utilizing newly emerging technology.

The mathematical basis and concepts of various algebras and systems of MVL are provided including comparisons among various systems. Early logic systems from the past century as well as more modern algebras are included.

The book also provides an examination of representations of multiple-valued logic suitable for implementation as data structures in automated computer applications including cube lists and maps. Decision diagram structures for multiple-valued applications are described in detail.

Dedicated chapters are provided for the discussion of reversible and quantum logic circuitry and representations. These topics have seen increasing interest in recent years by the computing community and directly relate to many topics in MVL. The quantum multiple-valued decision diagram is discussed in detail including algorithms for manipulation and minimization of the structure.

D. Michael Miller
Mitchell A. Thornton

CHAPTER 1

Multiple-Valued Logic Applications

1.1 INTRODUCTION

Multiple-Valued Logic (MVL) as discussed in this book is the study of discrete p-valued systems where $p > 2$, or in other words, nonbinary-valued systems. In the most general sense, both binary-valued (or digital) and discrete variables with an infinite number of values can be considered as MVL systems. In keeping with current convention, anytime we refer to an MVL system, we consider the system to utilize variables that can take on a discrete set of values with cardinality of three or more. Depending on the application, these systems can be considered to be mathematical structures, nonbinary discrete logic circuits, or symbolic logic constructs. In the context of this latter application, many logicians have referred to this topic as "many-valued logic" and this may well be a more precise name for the field since we are interested in systems that can evaluate many (meaning $p > 2$) different but distinct logic values in contrast to systems that can take on multiple logic values simultaneously. Likewise, the term "digital" has a general meaning of a set of integers or digits, but the colloquial meaning of a digital system is one that is binary-valued based on the two-member set of $\{0, 1\}$ and we also will use the term "digital" interchangeably with a binary-valued system. However, particularly in applications of circuits and the underlying algebras describing them, MVL has generally been the more popular phrase and we will conform to this trend and use the name MVL henceforth in our writing.

MVL has been studied and is of interest to engineers involved in various aspects of computing for over 40 years. A yearly international symposium on MVL has been held since the first meeting in May 1971 in Buffalo, New York. Over the years, the MVL community participating in these symposia has included researchers in the areas of discrete mathematics, computer circuit and system design, electronic device structure and architecture, information processing, discrete system algorithms and data structures, and so on. In this book, we will mainly present material from the viewpoint of the mathematical description of MVL as it pertains to computer system logic and algorithms although the concepts discussed here should be easily applied to other applications.

To provide motivation for learning about MVL, examples of some applications of MVL are provided here that generally fall under the category of computer system design and implementation. Specifically, we survey MVL uses in the areas of electronic design automation-computer aided design (EDA-CAD) and in circuit design. The use of MVL in these applications has not resulted in a singularly main approach, but certain aspects of these examples have benefited greatly through the use of MVL concepts, and it is our belief that MVL will continue to be an important tool for further advances in the application areas surveyed in this section.

In terms of circuit design, it is important to note that MVL methods are not germane only to implementations that encode logic values as more than two discrete voltage or current signals, but that MVL methods are also important as models for the initial design of logic circuits whether they are implemented with binary or MVL signal levels. The issue of whether the ultimate realization of a logic circuit is binary or not depends on the underlying technology and is independent of the use of MVL. As an example, complementary metal oxide semiconductor(CMOS)-based logic circuitry is generally implemented in binary logic since technology issues make binary the best choice. The point here is that this is merely an issue of the *encoding* of logic values and, as the following circuit examples indicate, the use of MVL concepts in the design stage often leads to circuits that exhibit better characteristics than would be obtained if only binary-valued logic were utilized.

1.2   MVL IN EDA-CAD METHODS

Modern integrated circuits (ICs) are so complex in terms of the number of basic components in which they are composed that it would be impossible to design and implement these devices without the use of CAD programs for EDA. Due to the properties of the basic components (i.e., transistors) used in the implementation of modern ICs, the most common type of logic used for actual circuit implementation by far is that of the digital logic. However, a large majority of the underlying algorithms used for the simulation, synthesis, and verification of digital ICs are based on MVL-based principles and methods.

MVL has a long history of use in EDA-CAD tools and the examples described in this section include hardware description languages (HDLs) for digital circuit simulation and synthesis, the tabular method for switching function minimization, the D-calculus as applied to digital circuit testing, and the symbolic trajectory evaluation (STE) technique for formal design verification to illustrate the diverseness of the meaning of different MVL systems in such CAD tools. While the meaning of these various MVL systems differs among the tools described, the common aspect is that MVL is extremely useful in EDA-CAD tools and is generally present in some form in almost all underlying CAD tool's algorithms.

1.2.1   Hardware Description Languages

HDLs are the current state-of-the-art medium for the description of digital circuits in the specification, simulation, synthesis, and verification/validation stages of the design cycle. The two most widely used HDLs are the languages VHDL (VHSIC hardware description language) and Verilog. Both of these HDLs utilize MVL systems internally.

TABLE 1.1: IEEE standard 1164 MVL values for VHDL

SYMBOL	NAME	MEANING
U	Unintialized	Initial default value for literal
X	Forcing Unknown	Forced unresolved logic value
0	Forcing 0	Forced logic value 0
1	Forcing 1	Forced logic value 1
Z	High Impedance	Logic value of open circuit
W	Weak Unknown	Weak unresolved logic value
L	Weak 0	Weak logic value 0
H	Weak 1	Weak logic value 1
-	Don't Care	Used for synthesis

1.2.1.1 VHDL. The VHDL language was first standardized by the Institute of Electrical and Electronics Engineers (IEEE) in 1987 as standard 1076. The original intent of VHDL was to provide a formal language for the description and specification of highly complex very high speed integrated circuits (VHSICs). In fact, VHDL is an acronym based upon the VHSIC acronym since it stands for VHSIC hardware description language. When VHDL was developed, a binary data type was specified that allowed for two-valued logic values of {0, 1}. In the years that followed, the development of VHDL, EDA simulation, and synthesis tools began to emerge from various CAD tool vendors that utilized VHDL as an input description medium. One of the difficulties in adopting the use of VHDL as a simulation or synthesis tool input mechanism was the fact that only the binary data type was available for describing internal circuit net values, thus states such as uninitialized net values or high-impedance state values were not easily represented. Commercial EDA vendors supplying automated digital circuit synthesis tools addressed this deficit by providing data type add-ons to the VHDL language that included these additional net values. Unfortunately, these MVL extensions were specific to each simulation or synthesis tool resulting in an overall lack of interoperability among tools developed by different vendors.

To address the interoperability issue among VHDL EDA-CAD tools, the IEEE formed a committee that developed standard 1164 in 1993, which provided for a common MVL extension for VHDL that can be utilized by EDA-CAD tool developers. Standard 1164 specified a nine-valued logic that is used almost universally by all CAD tools based on VHDL. The nine-valued MVL specification in standard 1164 is summarized in Figure 1.1.

The nine logic values specified in standard 1164 were chosen based on the underlying transistor technology used to build and fabricate digital circuitry. Specifically, the nine-valued logic allowed

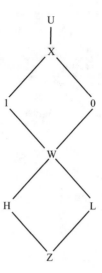

FIGURE 1.1: Resolution relationship of the nine-valued logic in IEEE standard 1164.

for a resolution relation to be specified that can be described as a lattice relation among the logic values. The resolution relation in VHDL allows for modeling a single net that has multiple drivers. When multiple drivers are present, it is necessary to resolve the signal to a single logic value. This allows for modeling circuits with "wired gates," pull-up and pull-down networks, and tri-state driver networks. The resolution relation graph is depicted in Figure 1.1.

It is emphasized that the nine-valued logic specified in standard 1164 is for the purpose of modeling binary-valued systems and not for modeling an MVL system. If it is desired to model MVL systems, where $p > 2$ in VHDL, a custom package analogous to the nine-valued logic of standard 1164 would be required, which contained significantly more than nine specific logic values. The actual number of logic values would depend upon the underlying technology of the MVL circuit being modeled [91].

1.2.1.2 Verilog. The Verilog HDL was developed in 1984 shortly after VHDL. Unlike VHDL, the original intent of Verilog was to serve as a hardware modeling language to be used as input to CAD simulator tools. "Verilog" is a shorthand word coined from "Verification of Logic." Due to the increasing complexity of digital ICs, a need was identified to be able to simulate circuits before detailed design to verify functional correctness. For Verilog to be utilized in a consistent manner among various EDA-CAD tool developers, the IEEE standardized the language in 1995 in standard 1364. Since simulation was an initial consideration during the design of the Verilog language, MVL needs were identified during the language design and a four-valued logic was included as part of the initial design of the language. Figure 1.2 gives a summary description of the MVL as specified in the Verilog IEEE standard 1364.

TABLE 1.2: IEEE standard 1364 MVL values for Verilog

SYMBOL	MEANING
0	Logic zero
1	Logic one
Z	High-impedance state
X	Unknown logic value

In contrast to the nine-valued logic of VHDL standard 1164 , the designers of the Verilog language chose the four-valued logic as specified in Table 1.2 and chose to allow certain primitive circuit elements to have a `strength` attribute. The combination of the four-valued logic values and the associated strength of the circuit element driving a net in a circuit simulation effectively provides a resolution relation similar to that in the nine-valued logic of standard 1164. The strength attribute values as defined in IEEE standard 1364-1995 allow certain gate types to have a `strength1` attribute of {`supply1`, `strong1`, `pull1`, `weak1`, `highz1`} and a `strength0` value of {`supply0`, `strong0`, `pull0`, `weak0`, `highz0`}.

1.2.2   Logic Synthesis

MVL is also used in many EDA-CAD automated synthesis techniques. When the tabular method commonly referred to as the "Quine-McCluskey" method was developed to minimize binary sum-of-products (SOP) expressions efficiently when implemented as a computer algorithm, it was necessary to allow for the inclusion of a "don't-care" value, X, in addition to the logic values of {0, 1}. In effect, a three-valued or ternary logic system of {0, 1, X} was necessary for implementation of the tabular method.

The tabular method [66,85] is a technique that was formulated to minimize the number of product terms in a SOP form for a binary-valued switching function. A common situation in the logic synthesis problem is the minimization of a multioutput binary switching function. The tabular method works well for single output switching functions and is essentially an embodiment of an algorithm for the solution of the general set covering problem [35]. Unfortunately, the application of the tabular method separately for each output of a multioutput binary switching function generally yields suboptimal results since each resulting output in SOP form depends on a minimal number of product terms, but no allowance is made to maximize product term sharing among the individual outputs. One solution is to enable term sharing of common products among the individual outputs to formulate the single output "characteristic function" representing the multioutput logic circuit. Once the characteristic function is formulated, the tabular method can be invoked and minimization is achieved whereby product term sharing is inherently accounted for.

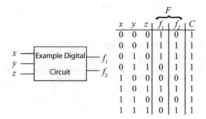

x	y	z	f_1	f_2	C
0	0	0	1	0	1
0	0	1	1	1	1
0	1	0	1	1	1
0	1	1	0	1	1
1	0	0	0	0	1
1	0	1	1	1	1
1	1	0	0	0	1
1	1	1	0	1	1

FIGURE 1.2: A two-output digital circuit and corresponding truth table.

The characteristic function of a multioutput digital logic circuit can be formulated as a binary-valued function with a MVL dependent variable set. As an example, consider the digital circuit characterized by the truth table shown in Figure 1.2. In the truth table, the two binary-valued outputs are labeled as f_1 and f_2 with the binary-valued dependent variable set $\{x, y, z\}$. The characteristic function is represented by C. With this point of view, the truth table becomes a list of product terms in the on-set of C and the $f_1 f_2$ terms are considered to be a single four-valued dependent variable of C. The resulting characteristic function C is binary-valued and depends on three binary-valued dependent variables $\{x, y, z\}$ and the single four-valued variable, F, whose polarity is given by the encoding provided by the $f_1 f_2$ pair. In algebraic sum-of-minterms form, the characteristic function is expressed as:

$$C = x^{\{0\}} y^{\{0\}} z^{\{0\}} F^{\{2\}} + x^{\{0\}} y^{\{0\}} z^{\{1\}} F^{\{3\}} + x^{\{0\}} y^{\{1\}} z^{\{0\}} F^{\{3\}}$$
$$+ x^{\{0\}} y^{\{1\}} z^{\{1\}} F^{\{1\}} + x^{\{1\}} y^{\{0\}} z^{\{0\}} F^{\{0\}} + x^{\{1\}} y^{\{0\}} z^{\{1\}} F^{\{3\}}$$
$$+ x^{\{1\}} y^{\{1\}} z^{\{0\}} F^{\{0\}} + x^{\{1\}} y^{\{1\}} z^{\{1\}} F^{\{1\}}$$

The characteristic equation can then be minimized in SOP form to yield the fewest product terms that C depends upon, thus inherently providing product term sharing during the minimization process. In converting back to digital form, the shared product terms correspond to those containing the $F^{\{3\}}$ literal while those that support only f_1 correspond to those containing $F^{\{2\}}$, and those supporting f_2 correspond to those containing $F^{\{1\}}$. Resulting terms that contain the literal $F^{\{0\}}$ are not included in the support set of either f_1 or f_2. More details on the extension of the tabular method utilizing the characteristic equation are provided in [97]. It is also noted that this technique can be used in minimization techniques other than the tabular method.

1.2.3   Logic Simulation

In addition to the simulation capabilities of the VHDL and Verilog HDLs, many EDA simulators have been developed that model circuits at different abstraction levels and for different purposes.

TABLE 1.3: Various digital logic simulators that use MVL

NAME	MVL VALUES	YEAR	REFERENCE
TEGAS	$\{0, 1, X, U, D, E\}$	1972	[40,108,109]
MOSSIM	$\{0, 1, X\}$	1981	[12]
LSim	$\{0, 1, X, Z, R, F, T\}$	1986	[19]
COSMOS	$\{0, 1, X\}$	1987	[15]
IRSIM	$\{0, 1, X\}$	1989	[93]
TRANALYZE	$\{0, 1, X, Z\}$	1991	[14]
VOSS	$\{0, 1, X, T\}$	1993	[99]
HALOTIS	$\{0, 1, R, F\}$	2001	[116]
Silos	$\{0, 1, X, Z\}$	2002	[14]
ModelSim	$\{0, 1, X, Z, U, W, L, H, -\}$	2003	[98]

A detailed survey of digital circuit simulators is presented in [36]. Many of these simulators utilize MVL internally to represent certain states or timing modes of the underlying digital devices.

One of the first digital circuit simulators that used MVL to represent timing parameters was the test generation and simulation (TEGAS) system [40,108,109]. The TEGAS simulator utilized a six-valued MVL set of $\{0, 1, X, U, D, E\}$ where $\{0, 1\}$ represent steady-state logic values, X represents an unknown value used for simulator initialization, U represents a signal in transition from low to high, D represents a signal in transition from high to low, and E represents a transient behavior such as a potential spike or hazard. TEGAS is a good example of how MVL principles can be used to model parameters of systems that do not just correspond to implemented logic levels, but in this case, are used to model discrete regions of timing behavior.

Many other digital circuit simulators have been developed in the past that use MVL internally. A summary of several of these appears in Table 1.3. The simulators listed in Table 1.3 vary in terms of the abstraction levels they represent and their individual capabilities. Some of the simulators support fault modeling while others strictly focus on switch or behavioral level functional simulation. Furthermore, although some of the logic values are indicated with similar symbols, the actual interpretation of each logic value may differ among the simulators. The reader is referred to the references for details on particular simulation programs.

1.2.4   Digital Hardware Testing

The testing problem is concerned with finding defects in a manufactured integrated circuit. During the fabrication of ICs, there is some percentage of devices that have errors (or faults) introduced during the production cycle. There is a large body of work developed in the past to characterize or model defects and to efficiently detect them. In support of these techniques, various MVL algebras have been developed for the purpose of generating a subset of input assignment values that are applied to manufactured devices for the purpose of determining if a fault is present in an IC. One of the first and simplest of such methods is the development of the D-calculus algebra [89] which, for our purposes, can be considered to be a four-valued algebra utilizing the logic values $\{0, 1, X, D\}$, where 0 and 1 are Boolean constants, X represents a don't-care condition, and D can take on either the value 0 or 1, but not both. When the output of the circuit is D or $\overline{D}$, the actual constant value observed is interpreted to indicate either the presence or nonpresence of a fault. When a suitable input assignment is applied to a device under test, the goal is to propagate the logic value D or $\overline{D}$ to the circuit output. If $D = 1$ or $\overline{D} = 0$, the device under test is free from the fault being tested whereas if $D = 0$ or $\overline{D} = 1$, the fault has been detected to be present. The determination of an appropriate set of circuit input assignments is, in general, a three-step process. These three steps are:

- Assign the value D (or $\overline{D}$) to an internal net where a stuck-at-1 or stuck-at-0 fault is assumed to be possibly present.

- Determine the appropriate net values required to propagate the value D or $\overline{D}$ to the output.

- On the basis of assigned internal net values from the previous step, propagate the internal net values backward in the circuit toward the inputs resulting in a corresponding test vector.

As an example consider the uppermost logic circuit in Figure 1.3, where the suspected fault is that the output of gate G2 always produces a logic-0 value (stuck-at-0) regardless of the input values. In step 1, this net is labeled with logic value D. In step 2, internal net values are assigned with values of $\{0, 1, X\}$ in order to propagate either D or $\overline{D}$ to the circuit output. In order to determine the appropriate internal net values, algebraic identities are used. Some of these identities, or D propagation rules are:

$$1 \wedge D = D$$
$$\overline{1 \wedge D} = \overline{D}$$
$$0 \vee D = D$$
$$\overline{0 \vee D} = \overline{D}$$
$$1 \oplus D = \overline{D}$$

After applying the D propagation rules, internal nets are annotated with appropriate logic values as shown in the middle circuit diagram in Figure 1.3. The third and final step of the D-algorithm requires propagating internally assigned net values backward toward the inputs in order to determine the test vector. Once again, appropriate algebraic identities are used based on the type of

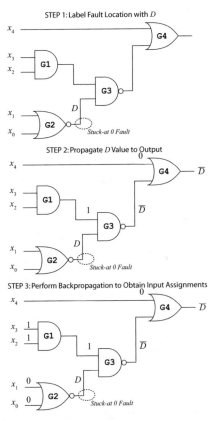

FIGURE 1.3: Example combinational circuit with D-algorithm applied.

gate present that allow for the determination of an appropriate gate input assignment. The bottom circuit in Fig. 3 shows the values assigned to the remaining nets yielding the resultant text vector of $\{x_4 = 0, x_3 = 1, x_2 = 1, x_1 = 0, x_0 = 0\}$.

This simple example illustrates how MVL is applied for test vector generation in the D-algorithm. Many other MVL algebras have been developed for the characterization and detection of more sophisticated faults such as delay faults and others. The determination of suitable fault models and associated algebras is a heavily researched topic.

1.2.5   Formal Verification

A recent EDA-CAD method for formal verification of design properties is the symbolic trajectory evaluation (STE) algorithm. While digital test algorithms are used to detect manufacturing errors after IC manufacture, verification techniques are used to uncover design errors or bugs before the circuit is manufactured. The STE implementation described in [100] utilizes a four-valued MVL system composed of the set $\{0, 1, X, T\}$, although the original idea for STE presented in [8] utilized

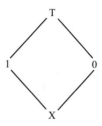

FIGURE 1.4: Resolution relationship of the four-valued MVL used in STE.

a ternary logic composed of $\{0, 1, X\}$. This MVL system is interesting since it represents a set of elements that have a resolution relationship as shown in Figure 1.4.

The idea behind the four-valued MVL system used in STE is that in some cases it is convenient in terms of simulation to symbolically represent the presence of both a logic-0 and a logic-1 in digital circuit, thus T contains both logic-0 and logic-1. The value X in this case represents an unknown logic value. STE can be considered to be a special case of model checking-based formal verification, which is a technique that allows properties of a design to be formally verified [22,67]. The properties are expressed in temporal logic, and the design under validation is modeled as a Kripke structure. In STE, the core algorithm used for ensuring that a property holds is the use of an internal logic simulator based on a set of logic values that include $\{0, 1, X, T\}$.

1.3   MVL CIRCUIT DESIGN

Although MVL methods are widely used in logic circuit design, the implementation of logic value encoding is mainly dependent on the target technology. The use of more than two signal levels to represent logic values is only one choice for MVL encoding. Circuits utilizing MOS transistors can support MVL by way of encoding the different logic levels as distinct binary strings. In this section, we will discuss examples of the use of MVL in circuit design using both signal level and binary string encoding.

1.3.1   Logic Circuit Design with MVL Signals

One of the most common reasons for considering the implementation of MVL circuits with more than two discrete signal levels is that of reducing wiring congestion as compared to binary-valued circuits. Using a single conductor to transmit three or more discrete voltage or current values allows for greater information content per wire and thus results in a circuit with fewer conductors than the binary-valued counterpart. In a survey paper by Hurst, [42], circuit cost models are formulated based on the number of signal lines and the range of values to be represented and it is concluded that a logic value of $p = 3$ would be more economical than the common value of $p = 2$.

In terms of voltage-mode circuits, relatively few MVL circuits have been commercially developed since the wiring congestion advantage is offset by the fact that most ICs produced in the last few decades are based on MOS transistors circuits that have increasingly smaller rail-to-rail voltage characteristics due to size reduction and the fact that such circuits are more power efficient in terms of static current dissipation when they are stabilized at a rail-voltage rather than an intermediate voltage representing another logic value. Small noise margins are another factor that prevent MVL voltage-mode circuits from being commercially produced on a large scale.

A considerable amount of work has been accomplished in using current levels to encode logic values. Such current-mode circuit implementations offer some advantages, one of which is the fact that an addition function is trivially implemented as a fan-in node since Kirchoff's current law states that all incoming currents to a circuit node sum together. Also, current-mode circuits have been promoted as being superior to voltage-mode circuits in terms of noise immunity. The basic building blocks of current-mode MVL circuits are usually considered to be the current mirror for fanout, scaling, and inversion and the use of current threshold detectors. Details of MOS transistor based MVL current-mode circuit design are provided in [32,46,105].

1.3.2   Memory Circuits

In contrast to current-mode circuits, a notable use of MVL in voltage-mode circuits is the nonvolatile flash memory device developed and marketed by Intel under the Strataflash™[31,114] product line. Each memory cell is based on a floating gate transistor where the polysilicon floating gate is charged to different levels. This results in the ability to store two bits per memory cell and, from an MVL point of view, four-valued logic is utilized with logic values represented internally as voltages. The bit pair "11" is represented by the erased state and bit pairs "01", "10", and "00" correspond to three different voltage ranges stored on the floating gate.

1.3.3   Computer Arithmetic Circuits

Arithmetic operations can be resource intensive in digital computers and the need to streamline such operations is a goal of designers of high-performance computer hardware. The basis of most arithmetic logic circuits is the addition operation. Many different addition circuit architectures have been devised with one of the main goals being to decrease circuit delay due to the inherent carry propagation characteristic. In the work reported in [6], a number system based on a redundant digit set was proposed that allowed for carry-free addition resulting in the construction of addition circuits that exhibit constant delay regardless of the wordsize of the addend and augend. Using this approach, addition circuitry can be constructed where operations on each digit of the operands can be performed in parallel thus overcoming the effect of overall circuit delay as a function of the length of the carry digit propagation chain.

A redundant number system is one whereby a single value can be represented by more than one digit string. This characteristic occurs when the number of distinct digits in the digit set is greater than the value of the radix or number base. In the original work of [6], such systems were formulated for radices of three and more. Later, these ideas were extended to the binary number system by [38,51,110] resulting in the ability to construct carry-free binary adders. As an example, a binary redundant number system can be constructed that utilizes the digit set $\{\bar{1}, 0, 1\}$ (in this discussion $\bar{1}$ represents the value -1) with a base or radix value of 2. As an example of the redundancy provided by this system, the unity value $(+1)$ can be represented by either the digit string 01 or $1\bar{1}$ since the radix polynomial expansion of these two digit strings both evaluate to $+1$:

$$(0) \times 2^1 + (1) \times 2^0 = (1) \times 2^1 + (-1) \times 2^0 = +1$$

Carry-free addition can be accomplished using redundant number systems by noting that the production of a nonzero carry-out digit only occurs when the digit 1 occurs in the same radix place for the addend and augend, or, the digit $\bar{1}$ occurs in that position. By exploiting redundant representations, nonzero carry-out digits can be restricted to avoid arbitrarily long carry chains. Details of carry-free propagation are provided in the cited references. For our purposes, we note that redundant computer arithmetic logic circuitry can be considered to be an MVL system with each distinct digit represented as a logic value.

In the past, the approach of considering redundant arithmetic circuits in terms of MVL systems has been utilized both for the purposes of designing circuits where more than two voltage or current levels are used to represent different logic values and for the case where such systems are designed through the use of MVL descriptions followed by a binary encoding of each digit set member.

The basic theory behind the use of MVL for arithmetic circuit design is relevant regardless of the encoding of the logic values and is driven largely by the suitability of devices currently available. In the past decade, the characteristics of MOS transistors have caused the realization of arithmetic logic circuits to be largely restricted to binary switching circuits where the transistors operate in either the cutoff or saturation modes. We emphasize that MOS transistor based implementations do not indicate that the use of MVL for arithmetic logic circuit design is irrelevant, rather it simply results in logic levels being represented as binary encodings during the implementation phase. MVL techniques are highly relevant for arithmetic circuit design and as new devices emerge, it may well be the case that logic values are encoded as voltage or current values in contrast to binary encoding, which is the predominant form in use currently.

1.3.4   Programmable Logic Arrays (PLAs)

Programmable logic arrays (PLAs) are a general purpose logic architecture that is capable of realizing any arbitrary binary-valued switching function. These circuits allow designers to implement combinational logic in the form of SOP expressions where internal interconnections are specified to produce specific functions. Figure 1.5 is a diagram of an example PLA capable of realizing five independent logic functions in SOP form that each depend on one or more of the four signals labeled $\{x_3, x_2, x_1, x_0\}$ with as many as eight different product terms each. The array of binary inverters

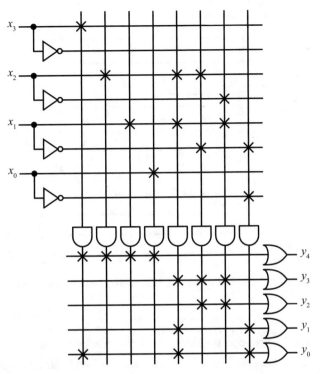

FIGURE 1.5: Example PLA supporting four inputs, eight product terms, and five outputs.

allows the input signals to be utilized in inverted or noninverted form. Traditionally, specified interconnections are illustrated by a multiplication symbol $\times$ present at the intersection of horizontal and vertical line intersections with the absence of a $\times$ indicating that no connection is present. In the example PLA in Figure 1.5, the following binary-valued logic functions are specified:

$$y_4 = x_3 + x_2 + x_1 + x_0$$
$$y_3 = x_2 x_1 + \overline{x}_2 x_1 + x_2 \overline{x}_1$$
$$y_2 = x_2 \overline{x}_1 + \overline{x}_2 x_1 = x_2 \oplus x_1$$
$$y_1 = x_2 x_1 + \overline{x}_1 \overline{x}_0$$
$$y_0 = x_3 + x_2 x_1 + \overline{x}_1 \overline{x}_0$$

An interesting application of MVL for PLA area reduction was developed in [94–96]. In this formulation, a standard PLA consisting of a programmable AND plane fanning out to a programmable OR plane is augmented with an array of active-low binary decoders called "literal decoders" preceding the AND plane instead of an array of inverters as is present in a standard PLA. An example of a small PLA including literal decoders is shown in Figure 1.6. The literal decoders

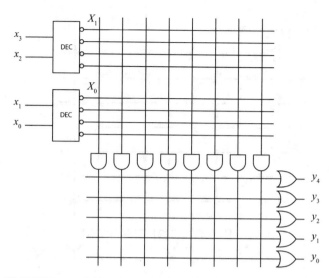

FIGURE 1.6: Example PLA augmented with literal decoders.

allow binary variables to be grouped together where the groups are defined as the set of variables present at the input of each decoder. The decoder outputs can then be interpreted as producing literals of a single MVL variable. In the example of Figure 1.6, 2:4 decoders are utilized. Thus, the decoder outputs represent a four-valued variable X_i in polarizations $\{\overline{X_i^{\{3\}}}, \overline{X_i^{\{2\}}}, \overline{X_i^{\{1\}}}, \overline{X_i^{\{0\}}}\}$. Specifically, the literal decoder in the top of Figure 1.6 generates various literals of the four-valued variable X_1, which has a binary encoding defined by the variable pair $x_3 x_2$. This configuration allows the functions to be implemented to be expressed and minimized using the four-valued dependent variable set $\{X_1, X_0\}$. In [94] several random functions were realized in standard PLAs and PLAs with literal decoders. A comparison of the resulting implementations showed that, on average, fewer product terms were required implying that the PLAs with literal decoders required fewer product terms (AND gates in the PLA) than would otherwise be needed when a standard PLA was used.

1.4   MVL CIRCUITS UTILIZING NEXT-GENERATION DEVICES

In addition to MVL circuits based on traditional mainstream devices such as MOS transistors, many newly emerging devices show promise for use in implementation of logic circuits that inherently support MVL. This section will discuss two such devices, resonant tunneling diodes (RTDs) and single electron transistors (SETs). Both of these devices exploit the quantum mechanical effect of tunneling and are constructed with internal "quantum islands" that support the quantum tunneling effect. As bias voltages increase, tunneling is enabled (or disabled) allowing for the current–voltage ($I\text{–}V$) characteristic curves to exhibit nonmonotonicity.

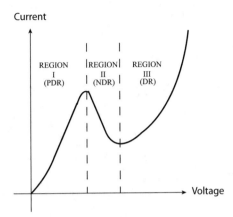

FIGURE 1.7: $I–V$ characteristic of resonant tunneling diode.

1.4.1   Resonant Tunneling Diodes

One of the more mature of the next-generation devices that exploit the quantum mechanical tunneling characteristic is the resonant tunneling diode (RTD). The $I–V$ relationship of an RTD contains three regions known as the positive differential resistive region (PDR), the negative differential region (NDR), and the diode region (DR). Figure 1.7 contains a plot of the $I–V$ characteristic of an RTD. The region of negative differential resistance occurs due to the quantum mechanical tunneling effect [18]. The tunneling effect is exploited through the incorporation of two different potential energy barriers in a two-terminal device and the effect of such structures was predicted in [20] and in more recent years, RTDs have been successfully fabricated that operate at room temperature.

The presence of two energy barriers in the RTD allows for a resonant energy level to occur based on the differences in the bandgap of the semiconductor materials used in the construction of the device. As a bias voltage is increased from 0 V across the device terminals, the RTD first operates in PDR Region I and the current flow through the device is approximately linear with respect to the applied voltage due to the emitted electrons reaching the same resonance energy levels as that present due to the two potential barriers allowing tunneling to occur. At some point, the voltage increases to a value referred to as the resonant voltage such that the energy of the electrons exceeds the resonant energy level of the barrier causing tunneling to be inhibited. As the applied voltage continues to increase beyond the resonant voltage level, the RTD enters the NDR Region II and current drops until the valley voltage is reached. In the NDR region, the current flow is approximately linear (with negative slope) for increasing voltage. When the valley voltage is exceeded, the RTD enters Region III and behaves like a normal diode.

Logic circuits based on RTDs exhibit desirable features in terms of high-speed operation, reduced circuit complexity, and low-power operation. The MVL aspect of RTD based circuits occurs

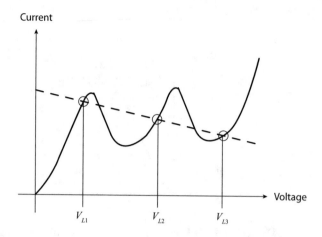

FIGURE 1.8: *I–V* characteristic of series RTD circuit.

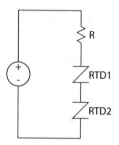

FIGURE 1.9: Example RTD series circuit schematic.

when two or more devices are interconnected with a resistance. Figure 1.8 illustrates the composite *I–V* characteristic obtained when two RTDs are connected in series with a resistance as shown in the schematic in Figure 1.9. The circled points in the plot of Figure 1.8 are intersections with the load line of the resistance (shown as a dotted line) and the series RTD *I–V* characteristic. These circled points represent differing stable voltage levels, V_{L1}, V_{L2}, and V_{L3} that can be used to encode logic values in an MVL circuit for a constant current.

Many different logic circuits have been designed such as MVL literal gates [121], programmable logic arrays [37], and MVL decoders [113] among others. A survey of the use of RTDs in MVL circuits is provided in [64].

1.4.2   Single Electron Transistors

SETs are three terminal devices that are constructed with two regions of dissimilar semiconductor interfaces that provide potential energy barriers similar to the RTD. The third terminal is connected to the "quantum island" occurring between the energy barriers and allows for a gate voltage to be

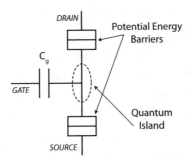

FIGURE 1.10: Single electron transistor diagram.

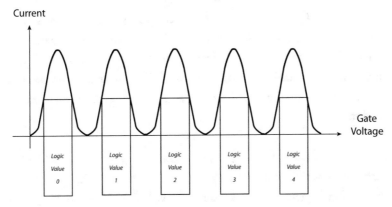

FIGURE 1.11: $I-V$ relationship for a SET with discrete logic levels shown.

applied to control electron flow across the island [55,112]. A diagram of the structure of an SET is given in Figure 1.10.

The characteristic curve for drain current versus gate voltage is nonmonotonic and approximately periodic yielding discrete voltage regions that may be used to encode logic values in an MVL circuit. Figure 1.11 contains an approximate $I-V$ relationship for a SET.

SET circuits have been designed for a variety of logic functions. Typically, these circuits utilize SETs in conjunction with more traditional devices such as MOS transistors yielding compact gates. As an example, a unary periodic literal gate is reported in [26] consisting of a MOS transistor, an SET, and a current source. Other logic circuits include counters and multipliers. In [45] SET circuits are described for analog-to-digital converters and multiple-valued (MV) addition circuits and in [111], a memory cell and MV multiplexer are reported.

1.4.3   MV Quantum Logic

An emerging field that has a direct relationship with MVL is that of quantum computing. The basic principle behind quantum computing is information processing using matter at the molecular, atomic, or particle level that exploits the characteristics of quantum state superposition and entanglement leading to the notion of "quantum parallelism." Quantum computing can be considered in the context of algorithms that exploit the quantum mechanical effects or, equivalently, the implementation of quantum logic circuits based on primitive elements that utilize the quantum mechanical nature of matter. The identification of quantum algorithms that allow for the solution of problems much more efficiently than that afforded by classical computers based on the Turing machine model have generated great interest in the computing community among theoreticians, algorithm developers, and logic designers. Although practical quantum computers do not yet exist, quantum devices and small quantum computation circuits have been demonstrated in various research laboratories.

While much of the effort currently devoted to quantum logic circuit and algorithm development is based on the use of Boolean (binary) observerables of quantum systems, there are numerous opportunities for the development of observerables with more than two eigenstates and the theory of MVL is very naturally applicable to the design and analysis of such systems. In past recent MVL symposia, research results concerning the application of MVL to quantum circuit design and analysis have been increasing in number and more researchers are devoting their investigations to this emerging application.

1.5   ORGANIZATION OF THE BOOK

This book will provide an overview of MVL algebras in Chapter 2. Although there have been several volumes published devoted solely to MVL algebra such as [17,30,79,87,88,124] and others, our goal in Chapter 2 is to provide an overview of the most commonly used algebras and most importantly to present them using a consistent notation. In providing a common and consistent notation for various MVL algebras, the reader should be able to more easily compare and contrast the strengths of each of the algebras and choose the one most appropriate to their particular application area.

In Chapter 3, we will discuss representations and data structures used to describe and manipulate MVL systems with a particular emphasis on decision diagram (DD) based techniques. Chapter 3 will first provide a brief review of truth tables, maps, cube lists, and their manipulation followed by a discussion of binary DD theory and efficient implementation software packages for DD representation and manipulation. Next, a discussion of the use of DD structures for the representation of MVL systems is given including aspects of the design of a MVL-DD (MDD) software package.

In Chapter 4, we will present material on reversible and quantum logic circuits. Such circuits are modeled with bijective logic functions and have many unique properties in the MVL case. Several examples of MVL reversible circuits are given and matrix methods for their representation is described.

Chapter 5 will give details of the quantum multiple-valued DD data structure. This DD has many unique characteristics that differ from the DDs described in Chapter 3 due mainly to the fact that the internal decomposition rule is quite different and that the structures are designed to represent unitary matrices. The discussion will include topics on the manipulation of QMDDs and details of the software package that supports them.

While the topics of Chapters 2–5 are by no means a comprehensive set of MVL topics, we believe these two subjects are well suited for an introduction of MVL and represent a good blend of past background and current ideas in the MVL community. Furthermore, we believe these two chapters are fairly independent in terms of MVL applications and our hope is that MVL practitioners with varying application interests will find these to be applicable. In the concluding Chapter 6, we summarize the topics presented in this book.

CHAPTER 2

MVL Concepts and Algebra

2.1 INTRODUCTION

This chapter is intended to survey algebraic concepts to enable the reader to quickly gain an introductory understanding of multiple-valued logic (MVL) fundamentals. More comprehensive texts are available that are devoted wholly or in part to algebraic concepts [30,79]. Basic concepts for the formal mathematical underpinnings of MVL systems are included. Although the intent is not to provide a comprehensive survey of discrete mathematics, a brief review of pertinent basic ideas is presented in order to provide a framework for the definitions of various algebras that are commonly used in the MVL literature. The chapter begins with a review of concepts in set theory with emphasis on set relations and their visualization as Hasse diagrams. Next, an overview of various structures such as fields, rings, and algebras is presented. The important property of functionally complete algebras is discussed. Functionally complete algebras are necessary in order to have the ability to describe all possible functions and their manipulation within a MVL system. This chapter concludes with a discussion of several specific logic systems and associated algebras that have been defined for MVL systems.

2.2 SETS AND RELATIONS

In this section it is assumed that the reader is familiar with basic set theory and operations among sets. We are interested in set theory since, in the context of MVL, it is convenient and common to describe logic system values as members of a set. Consider two sets A and B; the set $R \subseteq A \times B$ contains ordered elements of the form (a_i, b_j) and is said to express a relation among A and B. Relations are useful for expressing particular associations between the elements of two sets A and B based upon some property or two-place operator. It is often the case that a relation is defined among elements within the same set A in which case the elements of R are ordered pairs of the form (a_i, a_j), where $a_i, a_j \in A$.

Example 2.1. *Consider the set $A = \{1, 2, 3\}$. A relation may be formed based on the operator $>$ expressing those ordered pairs where a_i is strictly greater than a_j. In this case, $R = \{(3, 2), (3, 1), (2, 1)\}$. Notationally, R may be specified explicitly as $R = \{(3, 2), (3, 1), (2, 1)\}$ or as $R = \{A, >\}$.*

A particular element of R is often denoted as $a_i R a_j$ indicating the ordered pair (a_i, a_j) obeys the operator specified in the definition of the set R. Because relations can be defined based on two-place operators, they are also sometimes referred to as *binary relations*; however, we will simply use the term relation to avoid confusion with binary-valued logic systems. Set relations can be classified based on various properties they may hold. Some important properties of relations follow.

- When $a_i R a_i$ holds for all $a_i \in A$, then the relation R is said to be *reflexive*.
- When $a_i R a_j$ holds for all $a_i, a_j \in A$, then the relation R is said to be *symmetric* if it is also the case that $a_j R a_i$.
- When $a_i R a_j$ and $a_j R a_i$ hold for all $a_i, a_j \in A$, then the relation R is said to be *antisymmetric* if it is also the case that $a_i = a_j$.
- When $a_i R a_j$ and $a_j R a_k$ hold for all $a_i, a_j, a_k \in A$, then the relation R is said to be *transitive* if it is also the case that $a_i R a_k$.

Based on these properties, relations may be classified as *equivalence relations* or *partial order relations* according to the following definitions.

Definition 2.1.   A relation R is an **equivalence relation** if, and only if, the properties of reflexivity, symmetry, and transitivity hold.

Definition 2.2.   A relation R is said to be a **partial ordered relation** if, and only if, the properties of reflexivity, antisymmetry, and transitivity hold. Alternatively, such a relation R is also referred to as a **partially ordered set** or **poset**.

The general notation for a poset is $R = \{A, \leq\}$ where the $\leq$ operator represents any operator in conjunction with some A satisfying the definition of a poset. For a given poset, it can be the case that $a_i \leq a_j$ but $a_i \neq a_j$ resulting in $a_i < a_j$. When $a_i < a_j$ and there exists no $a_k \in A$ such that $a_i < a_k$ and $a_k < a_j$, then it is said that a_j *covers* a_i. This property leads to the following definition of a special class of posets.

Definition 2.3.   If either $a_i < a_j$, $a_i = a_j$, or $a_j < a_i$ for all pairs $(a_i, a_j) \in R$ where $\{A, \leq\}$ is a poset, then $\{A, \leq\}$ is said to be a **chain** or **linearly ordered poset**.

2.2.1   Relations as Graphs

A relation may be represented as a graph where vertices represent set elements and edges represent relations among those elements. If $a_i R a_j$, then the vertices representing a_i and a_j are adjacent and have a connecting graph edge. Features of a graph representation of a set relation allow for properties of the relation to be easily identified. For example, if a relation has the property of reflexivity, then

FIGURE 2.1: Directed graph representing example relation.

all vertices have edges that are self-loops. In general, graphs that represent set relations contain both directed and undirected edges. Directed edges are those edges that are characterized as originating from a source vertex and ending at a destination vertex. Traditionally, directed edges are denoted by drawing an arrowhead at the end of the edge at the destination vertex. If a set relation is symmetric, all graph edges are undirected since a_iRa_j and a_jRa_i; hence there is no notion of source and destination vertices. The property of transitivity is identified when, for all directed edges representing a_iRa_j and a_jRa_k, there is a corresponding edge representing a_iRa_k.

Example 2.2.   *The directed graph for the relation $R = \{A, >\} = \{(3, 2), (3, 1), (2, 1)\}$ over $A = \{1, 2, 3\}$ is shown in Figure 2.1.*

2.2.2   Hasse Diagrams

Hasse diagrams can be very helpful when analyzing algebras since they can be used to graphically visualize all possible functions the algebra can generate; particularly when the diagrams are sufficiently small. Hasse diagrams are representations of partially ordered sets, *posets*, which are graphs representing the relationships among the elements of the set. Hasse diagrams differ from the general digraph representation of set relations as described above in that edges corresponding to self-loops and transitive properties among the poset elements are suppressed. There is no loss in generality due to this edge suppression since Hasse diagrams are defined for posets only and the properties of reflexivity and transitivity must be present for a poset to exist. The set elements each uniquely correspond to a vertex and the directed edges depict the covering relationships among the elements. By definition, Hasse diagrams are drawn such that the *maximal element* vertices are drawn at the top of the graph and the *minimal element* vertices are at the bottom. For this reason, the arrowheads representing the order of the pairs in the poset are also suppressed since those vertices that are above others represent the covering relationship. Maximal element vertices have no incoming edges and only have outgoing edges whereas minimal element vertices only have incoming edges with no outgoing edges; thus the maximal elements appear at the top of the Hasse diagram and the minimal elements appear at the lowermost position.

2.2.3   Finite Lattices

Finite lattices are a subset of Hasse diagrams that have additional properties not necessary for the definition of a Hasse diagram. A *lattice* represents a particular type of poset where each pair of elements has a *minimal element* or least upper bound (LUB) and a *maximal element* or greatest lower bound (GLB). A lattice is said to be *complete* if every nonempty subset of A has a LUB and GLB. Every finite lattice, where A is a finite set, is complete. A complete lattice has a greatest element denoted as $\mathbf{1}$ and a least element denoted as $\mathbf{0}$. All finite lattices have a unique LUB and a unique GLB.

Lattices may be defined in algebraic terms, A complete lattice is a sextuple $\langle A, \leq, +, \cdot, \mathbf{0}, \mathbf{1} \rangle$, where the $+$ operator corresponds to the GLB, the $\cdot$ operator corresponds to the LUB, and the following properties are satisfied:

- For all $a_i, a_j \in A$, $a_i + a_j \in A$ and $a_i a_j \in A$ hold. This is the property of *closure*.
- For all $a_i, a_j, a_k \in A$, both $a_i + (a_j + a_k) = (a_i + a_j) + a_k$ and $a_i(a_j a_k) = (a_i a_j)a_k$ hold. This is the property of *associativity*.
- For all $a_i, a_k \in A$, $a_i + a_j = a_j + a_i$ and $a_i a_j = a_j a_i$ hold. This is the property of *commutativity*.
- The elements $\mathbf{0}$ and $\mathbf{1}$ have the properties that for each $a_i \in A$, $a_i + \mathbf{0} = a_i$, $a_i + \mathbf{1} = \mathbf{1}$, $a_i \mathbf{0} = \mathbf{0}$ and $a_i \mathbf{1} = a_i$. This is the property of existence of *additive* and *multiplicative identities*.
- For each $a_i \in A$, $a_i + a_i = a_i$ and $a_i a_i = a_i$. This is the property of *idempotence*.

Furthermore, a lattice is said to be *distributive* if

- for all $a_i, a_j, a_k \in A$, $a_i(a_k + a_j) = a_i a_k + a_i a_j$ and $a_i + a_j a_k = (a_i + a_j)(a_i + a_k)$.

A complete distributed lattice is *complemented* if

- for each $a_i \in A$, there is an element $a_i' \in A$ such that $a_i + a_i' = \mathbf{1}$ and $a_i a_i' = \mathbf{0}$.

The above seven axioms are invariant if $+$ and $\cdot$ are interchanged and $\mathbf{0}$ and $\mathbf{1}$ are interchanged. This is the principle of *duality*. Duality is important since any proposition regarding a complete lattice that is true is also true under interchange of $+$ and $\cdot$ and the interchange of $\mathbf{0}$ and $\mathbf{1}$.

Example 2.3.   *Consider the poset $R = \{A, P()\}$ where $A = \{a_i, a_j, a_k\}$ and $P()$ represents the operation of generating the power sets of an argument set. This is an instance of a complete lattice, $R = P(A)$, where each member of $P(A)$ is a subset of A and the definition of a poset is satisfied. The lattice diagram for this example is shown in Figure 2.2.*

Other examples of lattices are the resolution diagrams presented in Figs. 1.1 and 1.4 in Chapter 1 where the posets R are formed using the VHDL and STE logic values and their corresponding

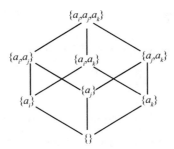

FIGURE 2.2: Lattice diagram of the example poset.

resolution functions. The examples in Chapter 1 illustrate how MVL can be used to formally define logic systems that are useful in EDA-CAD tools used for binary-valued circuit design.

2.3   ALGEBRAIC STRUCTURES

In preparation for surveying various MVL systems, this section reviews the fundamentals underpinning the algebraic structures to be considered. Some familiarity with elementary set theory and with the concepts of discrete mathematics and two-valued switching theory as normally taught in an undergraduate program in computer science or computer engineering is assumed.

2.3.1   Functions

A function f is a many-to-one mapping from a domain D to a range R denoted $f : D \longrightarrow R$. A function is totally specified if the mapping is defined for every element of D, otherwise the function is partially specified.

In this book, we shall be concerned with functions over finite spaces. Let E be a finite set of values with a cardinality of three or greater and let E^n denote the Cartesian product of n copies of E.

Definition 2.4. A multiple-valued function is one where $f : E^n \longrightarrow E$.

Definition 2.5. A binary Boolean (sometimes simply termed binary) function is one where $E = \{0, 1\}$. For clarity we shall replace E with $B = \{0, 1\}$ for the binary case giving $f : B^n \longrightarrow B$.

2.3.2   Rings

A *ring* is a quintuple $\langle E, +, \cdot, \mathbf{0}, \mathbf{1} \rangle$. E is a finite or infinite set which includes elements $\mathbf{0}$ and $\mathbf{1}$. There are 2 two-place operations $+$ and $\cdot$. The normal convention of omitting the $\cdot$ operator, *e.g.* writing ab in place of $a \cdot b$, will be used and $+$ takes precedence over $\cdot$.

A ring satisfies the following:

- For all a and b in E, $a + b$ and ab, are in E (closure over $+$ and $\cdot$).
- For all a, b, and c in E, both $a + (b + c) = (a + b) + c$ and $a(bc) = (ab)c$ hold (associativity property).
- For all a and b in E, $a + b = b + a$ (commutativity over $+$).
- The elements $\mathbf{0}$ and $\mathbf{1}$ have the properties that for each a in E, $a + \mathbf{0} = a$ and $a\mathbf{1} = \mathbf{1}a = a$ (additive and multiplicative identities exist).
- For each a in E, there is an element a' in E such that $a + a' = \mathbf{0}$ (additive complements exist).
- For all a, b, and c in E, $a(b + c) = ab + ac$ hold (distributivity over $\cdot$).

In addition to the above properties, if for all a and b in E it holds that $ab = ba$, the ring is called commutative or Abelian, otherwise the ring is non-Abelian.

A ring is a structure in which one can *add*, *subtract*, and *multiply* in a fashion very similar to what we know from elementary arithmetic. In fact, the integers under usual addition ($+$) and multiplication ($\cdot$) form an Abelian ring although in this case E is the infinite set of all integers.

2.3.3   Fields

A *field* is a ring where the $\cdot$ operation is commutative, and each element except $\mathbf{0}$ has a multiplicative inverse. A field satisfies the following:

- For all a and b in E, $a + b$ and ab, are in E.
- For all a, b, and c in E, both $a + (b + c) = (a + b) + c$ and $a(bc) = (ab)c$ hold.
- For all a and b in E, $a + b = b + a$ and $ab = ba$.
- The elements $\mathbf{0}$ and $\mathbf{1}$ have the property that for each a in E, $a + \mathbf{0} = a$ and $a\mathbf{1} = a$.
- For each a in E, there is an element a' in E such that $a + a' = \mathbf{0}$ and for each nonzero a in E, there is an element a'' in E such that $aa'' = \mathbf{1}$.
- For all a, b, and c in E, $a(b + c) = ab + ac$ holds.

One can add, subtract, multiply, and divide, except by $\mathbf{0}$, over a field. Comparing the properties given above, we see that every field is a ring. The converse does not necessarily hold.

Example 2.4.   *Given $+$ and $\cdot$ defined as*

+	0	1	2	3
0	0	1	2	3
1	1	2	3	0
2	2	3	0	1
3	3	0	1	2

·	0	1	2	3
0	0	0	0	0
1	0	1	2	3
2	0	2	0	2
3	0	3	2	1

$\langle \{0, 1, 2, 3\}, +, \cdot, \mathbf{0}, \mathbf{1} \rangle$ *is a ring. However, this is not a field since there is no b such that* $2b = \mathbf{1}$.

Example 2.5.   *Given* $+$ *and* $\cdot$ *defined as*

+	0	1	2	3
0	0	1	2	3
1	1	0	3	2
2	2	3	0	1
3	3	2	1	0

·	0	1	2	3
0	0	0	0	0
1	0	1	2	3
2	0	2	3	1
3	0	3	1	2

$\langle \{0, 1, 2, 3\}, +, \cdot, \mathbf{0}, \mathbf{1} \rangle$ *is a field.*

2.4   BOOLEAN ALGEBRA

Boolean algebra over the two-valued logic system using the set $\{0, 1\}$ is the theoretical underpinning for two-valued switching circuits. This is in fact one instance of a more general family of Boolean algebras. A good understanding of these algebras, and particularly their limitations, is essential to understanding the algebraic structures required for multiple-valued systems.

2.4.1   Basic Definition

A *Boolean algebra* is a sextuple $\langle B, +, \cdot,', \mathbf{0}, \mathbf{1} \rangle$ where B is a set of elements including $\mathbf{0}$ and $\mathbf{1}$, $+$ and $\cdot$ are two-place operations and $'$ is a one-place operation. The $\cdot$ operator takes precedence over $+$ and $'$ takes precedence over both. The $\cdot$ will typically be omitted so that, for example, $a \cdot b$ is written as ab. A Boolean algebra satisfies the following basic axioms:

- For a in B, a' is also in B.
- For a and b in B, $a + b$ and ab are in B.
- For all a and b in B, $a + b = b + a$ and $ab = ba$.
- For all a, b, and c in B, both $a(b + c) = ab + ac$ and $a + bc = (a + b)(a + c)$.
- For a in B, both $a\mathbf{1} = a$ and $a + \mathbf{0} = a$.

- For a in B, both $aa' = 0$ and $a + a' = 1$.
- $0 \neq 1$.

The properties below follow from the basic axioms of a above:

- The complement is unique and $(a')' = a$ for all a in B.
- For all a, b, and c in B, both $a + (b + c) = (a + b) + c$ and $a(bc) = (ab)c$.
- For all a in B, $a0 = 0$ and $a + 1 = 1$.
- For all a and b in B, both $(a + b)' = a'b'$ and $(ab)' = a' + b'$.
- For all a and b in B, both $a(a + b) = a$ and $a + ab = a$.
- For all a in B, both $aa = a$ and $a + a = a$.
- $0' = 1$ and $1' = 0$.

It is important to note that the principle of duality is present in a Boolean algebra. In comparing the above definition of a Boolean algebra to the lattice properties listed earlier, one finds that a Boolean algebra is a complemented, complete distributed lattice. It is also noted that the most popular Boolean algebra in use is that defined over a two-valued logic system. In general, a Boolean algebra may be defined over a logic system with any number of finite logic values, however only those Boolean algebras defined over $p = 2^n$ logic values are functionally complete.

2.4.2   Alternative Definitions

Consider the algebra defined by $\langle B, \overline{\wedge}, 1 \rangle$ with 1 in B and the following axioms:

- For all a and b in B, $a \overline{\wedge} b$ is in B.
- For notational convenience, let a^+ denote $a \overline{\wedge} a$. For all a in B, $(a^+)^+ = a$.
- For all a and b in B, $a \overline{\wedge} b = b \overline{\wedge} a$.
- For all a in B, $a^+ 1 = a$.
- For all a in B, $a^+ \overline{\wedge} a = 1$.
- For all a, b, and c in B, $a \overline{\wedge} (b \overline{\wedge} c) = ((a \overline{\wedge} b^+) \overline{\wedge} (a \overline{\wedge} c^+))^+$.
- There are at least two distinct elements in B.

Comparing this to the definition of a Boolean algebra in the previous section, we find the alternative definition here also defines a Boolean algebra with $a \overline{\wedge} b = (ab)'$. Note that $a \overline{\wedge} a = a^+ = a'$.

The operator $\overline{\wedge}$ is the well-known NAND (NOT AND) operation which is of major importance in two-valued systems because of the ease with which it is implemented in particular technologies. By duality, it is clear a similar definition of a Boolean algebra is possible using the NOR (NOT-OR) operation alone as $\langle B, \overline{\vee}, 0 \rangle$.

These two formulations of complete Boolean algebras are of interest since they only require a single two-place operation. The $\overline{\wedge}$ operator is also known as the Sheffer Stroke [101]. The $\overline{\vee}$ operator was discovered by Peirce [81] over 30 years earlier to similarly allow a functionally complete algebra to be formed with a single two-place operator. The Sheffer Stroke has been generalized to MVL systems where $p > 2$ and several two-place operators satisfy the definition of a Sheffer Stroke meaning that an MVL algebra can be defined with only one of these two-place operations [59,122]. Although it is not the case in this book, the symbols $|$ and $\uparrow$ are often used to represent the Sheffer Stroke operator while the $\downarrow$ is used to represent the operation of Peirce.

2.4.3   Boolean Normal Forms

SOP representations are a well-known concept from two-valued switching theory that is in fact applicable in any Boolean algebra.

Definition 2.6 Boolean Expressions.

(i) Any variable is a Boolean expression.

(ii) If P and Q are Boolean expressions, then so are $P + Q$, $P \cdot Q$ and P'. Brackets are of course allowed to show precedence of evaluation.

(iii) Nothing is a Boolean expression except as defined by (i) and (ii).

It is crucial to note that the expression representing a Boolean function is not unique. In fact, that is the basic challenge, to find an optimal expression for a given function, where optimal can take many different forms depending on the context. In circuit design for example, optimality can relate to speed, number of devices, testability, power consumption, or as is most often the case, some combination of these and other factors.

Theorem 2.1 Shannon Decomposition.   A Boolean expression P may be expressed in the form $P = x'P_0 + xP_1$ where x is a variable in P which does not appear in P_0 or P_1.

Proof.   The proof is by induction on the number k of occurrences of the operators $+, \cdot, '$ in P. For $k = 0$, either P is x, or P is some other variable y. If $P = x$, $P_0 = 0$, and $P_1 = 1$ so that $P = x'0 + x1 = x$. If $P = y$, $P_0 = P_1 = y$ so that $P = x'y + xy = y$.

Now assume the result for all $j \leq k$ and consider the case of $k + 1$ operators in P. By the construction of a Boolean expression (Definition 2.4.3), there is a single final operation. The three cases are considered in turn:

- For the case of $'$, we have $P = Q'$, where Q is an expression having k operators which by the inductive assumption can be written as $Q = x'Q_0 + xQ_1$ with x not appearing in

Q_0 or Q_1. Hence, $P = (x'Q_0 + xQ_1)' = (x'Q_0)'(xQ_1)' = (x + Q_0')(x' + Q_1') = x'Q_0' + xQ_1'$. Setting $P_0 = Q_0'$ and $P_1 = Q_1'$, we have $P = x'P_0 + xP_1$ so the case is proven.

- For the case of $+$, we have $P = Q + R$ where both R and Q involve at most k operators. Hence $P = x'Q_0 + xQ_1 + x'R_0 + xR_1 = x'(Q_0 + R_0) + x(Q_1 + R_1)$. Setting $P_0 = Q_0 + R_0$ and $P_1 = Q_1 + R_1$ yields $P = x'P_0 + xP_1$ so the case is proven.

- The case of the final operator being $\cdot$ is proven in a manner analogous to the $+$ case.

Application of Theorem 2.4.3 by induction on n yields the following theorem.

Theorem 2.2 Disjunctive Normal Form.

Any Boolean expression P of n variables $x_1, \ldots, x_n$ can be written in the form

$$P = \sum_{v=0}^{2^n-1} x_1^{E_1} x_2^{E_2} \cdots x_n^{E_n} \cdot m_v$$

where $(E_1, E_2, \ldots, E_n)$ is the n-digit binary expansion of v and in this context a superscript of 0 denotes complementation of the variable while a superscript of 1 denotes the variable appears uncomplemented. Each m_v is **0** or **1** and $\sum$ denotes that the terms are combined using $+$ operations.

For example, this theorem states that a Boolean expression of two variables x_1 and x_2 can be written as $x_1'x_2' \cdot m_0 + x_1x_2' \cdot m_1 + x_1'x_2 \cdot m_2 + x_1x_2 \cdot m_3$.

The reader will recognize the disjunctive normal form as a *standard SOP* or *minterm expansion*; those terms being more commonly used in switching theory. The duality of Boolean algebra leads to the following theorem which can again be proven by inductive application of Theorem 2.4.3. The conjunctive normal form is also termed a *standard product of sums* (POS) or *maxterm expansion*.

Theorem 2.3 Conjunctive Normal Form. Any Boolean expression P of n variables $x_1, \ldots, x_n$ can be written in the form:

$$P = \prod_{v=0}^{2^n-1} \left(x_1^{E_1} + x_2^{E_2} + \cdots + x_n^{E_n} + M_v \right)$$

Each M_v is 0 or 1 and $\prod$ denotes the terms are combined using $\cdot$ operations.

It is clear from the construction, that the disjunctive normal form and the conjunctive normal form are both unique. Any given function has only one representation of each form and no two distinct functions have a common normal form.

Definition 2.7. A function $f(x_1, \ldots, x_n)$ with domain E^n and range E is closed with respect to $\hat{E}$, a proper subset of E, if $f(x_1 \cdots x_n)$ is in $\hat{E}$ whenever all x_i are in $\hat{E}$.

Lemma. Any Boolean function is closed on $\{0, 1\}$.

Proof.   The proof follows from the fact any Boolean expression, and hence the corresponding function, can be expressed in disjunctive normal form. It follows immediately from the definition of that form that whenever all x_i are in $\{0, 1\}$, the resultant function value is also in $\{0, 1\}$.

Theorem 2.4.   In any Boolean algebra, there are exactly 2^{2^n} functions of n variables.

Proof.   This also follows directly from the disjunctive normal form since in the expression

$$P = \sum_{v=0}^{2^n-1} x_1^{E_1} x_2^{E_2} \cdots x_n^{E_n} \cdot m_v$$

each m_v is 0 or 1. There are clearly 2^{2^n} ways to assign values to the m_v each leading to a unique function.

This theorem shows the fundamental limitation of Boolean algebra in terms of dealing with MVL functions. Whereas a Boolean algebra can represent 2^{2^n} functions of n variables, as we have noted at the beginning of this chapter, there are p^{p^n} distinct functions of n p-valued variables.

2.5   LOGIC SYSTEMS AND ALGEBRAS

Since Boolean algebra is not in general adequate for multiple-valued functions due to its lack of completeness over sets of logic values that do not have a cardinality of 2^n, we must consider alternative algebraic structures. The purpose in developing an algebra is to provide for a concise well-defined framework for expressing and manipulating functions. In many applications, logic design in particular, a second and equally important consideration is that the operators of the algebra have simple and efficient circuit implementations. In this section, we present a variety of MVL systems and associated algebras.

In informal terms, the algebras we are interested in consist of a finite set of values including a pair of identity elements and a set of operators selected from the unary (one-place) and two-place functions defined over the set of elements. Note that the two-place functions can correctly be called binary functions, but we shall refer to them as two-place functions to avoid the obvious confusion arising from the terminology "binary MVL function."

Many MVL systems have been developed as extensions of classical two-valued propositional logic in order to address propositions not easily modeled with binary truth values. Much of this work was accomplished by researchers of symbolic logic and discrete mathematics. For the purposes of modeling MVL circuitry, or using MVL principles in the development of EDA-CAD tools, it is important to utilize MVL systems that can be formulated as complete algebras and to furthermore restrict such algebras to containing easily realizable and convenient one-place and two-place operators. We shall first consider MVL operators in general and then survey several MVL systems and their associated algebras.

The various MVL systems described here have been defined over finite sets of logic values with a cardinality of $p \geq 3$. Although infinite valued logic systems are certainly an important MVL

TABLE 2.1: Two-place MVL operators

NAME	NOTATION	DEFINITION	REFERENCE
Min	$x \cdot y$	x if $x < y$, y otherwise	Post 1921
Max	$x + y$	x if $x > y$, y otherwise	Post 1921
Mod-sum	$x \oplus y$	$(x + y) mod_p$	
Mod-difference	$x \ominus p$	$(x - y) mod_p$	
Truncated sum	$x +_t y$	$min(p - 1, sum(x, y))$	

topic, we concentrate on finite systems although most of the material presented here is applicable and easily extended to infinite-valued systems. In addition to the syntactical and formal definitions of these systems, the semantics of such systems is also included to provide background and an intuitive basis for the original meaning that motivated their development. As examples for these various logic systems, we will focus on three-valued implementations; however the logic systems presented may be easily generalized for the case where $p > 3$.

2.5.1   MVL Operators

In considering MVL algebras, we are most interested in those where the algebraic operators represent functions which have straightforward circuit implementations and which have sufficient representative power that effective circuit implementations can be constructed for general p-valued functions. There are p^{p^2} two-variable p-valued functions and p^p one-variable p-valued functions, so there is considerable choice for the algebraic operators that might be used.

Table 2.1 shows five two-place operators of particular interest. Table 2.2 defines these operators for $p = 2, 3, 4$ which are the values currently of most practical interest. The operators min, max and mod-sum, correspond to the familiar AND, OR and EXOR two-place functions, for systems where $p = 2$. The mod-difference function is the same as mod-sum for $p = 2$. The truncated sum function is mathematically the same as the OR function for systems where $p = 2$. In general there exist p^{p^2} different two-place functions that can be assigned unique operators for a p-valued logic system. In the binary-valued case, this amounts to an easily handled 16 different functions, however the total number of two-place functions quickly becomes unwieldy for larger values of p where 16,683 different functions exist for $p = 3$ ternary systems and 4,294,967,269 different functions exist for $p = 4$ MVL systems.

From the definitions of the min and max operators, it is readily verified that the properties given in Table 2.3 hold. The reader will recognize these as direct generalizations of well-known properties of two-valued Boolean algebra.

TABLE 2.2: Two-place MVL operators for $p = 2, 3, 4$

	$P = 2$		$P = 3$		$P = 4$	

Min

$\cdot$	0	1
0	0	0
1	0	1

$\cdot$	0	1	2
0	0	0	0
1	0	1	1
2	0	1	2

$\cdot$	0	1	2	3
0	0	0	0	0
1	0	1	1	1
2	0	1	2	2
3	0	1	2	3

Max

$+$	0	1
0	0	1
1	1	1

$+$	0	1	2
0	0	1	2
1	1	1	2
2	2	2	2

$+$	0	1	2	3
0	0	1	2	3
1	1	1	2	3
2	2	2	2	3
3	3	3	3	3

Mod-sum

$\oplus$	0	1
0	0	1
1	1	0

$\oplus$	0	1	2
0	0	1	2
1	1	2	0
2	2	0	1

$\oplus$	0	1	2	3
0	0	1	2	3
1	1	2	3	0
2	2	3	0	1
3	3	0	1	2

Mod-difference

$\ominus$	0	1
0	0	1
1	1	0

$\ominus$	0	1	2
0	0	2	1
1	1	0	2
2	2	1	0

$\ominus$	0	1	2	3
0	0	3	2	1
1	1	0	3	2
2	2	1	0	3
3	3	2	1	0

Truncated sum

$+_t$	0	1
0	0	1
1	1	1

$+_t$	0	1	2
0	0	1	2
1	1	2	2
2	2	2	2

$+_t$	0	1	2	3
0	0	1	2	3
1	1	2	3	3
2	2	3	3	3
3	3	3	3	3

The number of one-place, or unary, functions for a p-valued logic system is p^p. For $p = 2$, there are only four unary functions, the identity, negation, and the two constant functions 0 and 1. There is thus really no choice to be made as these four functions are all reasonably easy to implement in circuitry or in software and are small in number. However, for ternary systems where $p = 3$, there

TABLE 2.3: Properties of the min and max operators

Idempotent	$x + x = x$	$x \cdot x = x$
Commutative	$x + y = y + x$	$x \cdot y = y \cdot x$
Associative	$(x + y) + z = x + (y + z)$	$(x \cdot y) \cdot z = x \cdot (y \cdot z)$
Absorption	$x + x \cdot y = x$	$x \cdot (x + y) = x$
Distributive	$x + y \cdot z = (x + y) \cdot (x + z)$	$x \cdot (y + z) = x \cdot y + x \cdot z$
Null element	$x + 0 = x$	$x \cdot 0 = 0$
Universal element	$x + (p - 1) = p - 1$	$x \cdot (p - 1) = x$

are 27 unary functions, which, less the identity and the p constant functions, leaves 23 candidates for unary operators. For $p = 4$, there are 256 candidate one-place functions in total. Those one-place functions that have been commonly used in past literature and which we will concentrate on are shown in Table 2.4.

A word about notation is in order. There has been fairly strong consensus on the notation for MVL two-place operators, although certainly not unanimity. In contrast, a wide variety of notations have been used for unary operators. Here we have chosen notation with regard for past practice and also in an attempt for clarity. Table 2.5 shows unary operators for $p = 2, 3, 4$. The table does not include the constant functions nor operators that are equivalent to constants (*e.g.*, $x^{[0, p-1]} = p - 1$). We also note that $C_a(x) = x^{\{a\}} = x^{[a]}$ so only $C_a(x)$ is shown. Furthermore, many of the one-place operators may be defined in terms of one another. As an example, the threshold functions may be defined in terms of the window functions as $x^{\underline{a}} = x^{[a, p-1]}$.

TABLE 2.4: Unary MVL operators

NAME	NOTATION	DEFINITION	REFERENCE
Cycle	x^k	$(x + k) mod_p$	Post 1921
Successor	$\overrightarrow{x}$	$(x + 1) mod_p$	
Predecessor	$\overleftarrow{x}$	$(x - 1) mod_p$	
Negation	$\overline{x}$	$(p - 1) - x$	Łukasiewicz 1920
Decisive literal	$C_a(x)$	$p - 1$ if $x = a$, 0 otherwise	Epstein 1960
Window literal	$x^{[a,b]}$	$p - 1$ if $a \leq x \leq b$, 0 otherwise	Allen and Givone 1968
Selection literal	x^S	$p - 1$ if $a \in S$, 0 otherwise	Allen and Givone 1968
Threshold literal	$x^{\underline{a}}$	$p - 1$ if $x \geq a$, 0 otherwise	Birk and Farmer 1974

TABLE 2.5: Unary MVL operators for $p = 2, 3, 4$ (selection literals omitted for $n = 4$)

$p = 2$

x	$\bar{x}$
0	1
1	0

$p = 3$

x	x^1	x^2	$\overrightarrow{x}$	$\overleftarrow{x}$	$\bar{x}$	$C_0(x)$	$C_1(x)$	$C_2(x)$	$x^{[0,1]}$	$x^{[1,2]}$
0	1	2	1	2	2	2	0	0	2	0
1	2	0	2	0	1	0	2	0	2	2
2	0	1	0	1	0	0	0	2	0	2

x	$x^{\{0,1\}}$	$x^{\{0,2\}}$	$x^{\{1,2\}}$	$x^{\underline{1}}$	$x^{\underline{2}}$
0	2	2	0	0	0
1	2	0	2	2	0
2	0	2	2	2	2

$p = 4$

x	x^1	x^2	x^3	$\overrightarrow{x}$	$\overleftarrow{x}$	$\bar{x}$	$C_0(x)$	$C_1(x)$	$C_2(x)$	$C_3(x)$
0	1	2	3	1	3	3	3	0	0	0
1	2	3	0	2	0	2	0	3	0	0
2	3	0	1	3	1	1	0	0	3	0
3	0	1	2	0	2	0	0	0	0	3

x	$x^{\{0,1\}}$	$x^{\{0,2\}}$	$x^{\{1,2\}}$	$x^{\{0,1,2\}}$	$x^{\{0,3\}}$	$x^{\{1,3\}}$	$x^{\{0,1,3\}}$	$x^{\{2,3\}}$
0	3	3	0	3	3	0	3	0
1	3	0	3	3	0	3	3	0
2	0	3	3	3	0	0	0	3
3	0	0	0	0	3	3	3	3

x	$x^{\{0,2,3\}}$	$x^{\{1,2,3\}}$	$x^{[0,1]}$	$x^{[0,2]}$	$x^{[1,2]}$	$x^{[1,3]}$	$x^{[2,3]}$	$x^{\underline{1}}$	$x^{\underline{2}}$	$x^{\underline{3}}$
0	3	0	3	3	0	0	0	0	0	0
1	0	3	3	3	3	3	0	3	0	0
2	3	3	0	3	3	3	3	3	3	0
3	3	3	0	0	0	3	3	3	3	3

It is important to note that the unary operators are not unique, in that certain unary functions have different operator specifications. Predecessor and successor are clearly particular cases of the more general cycle operation. They have historically been considered as distinct since they have direct circuit implementations and need not be implemented as a general cycle circuit. Other equivalences exist amongst the window, selection, and threshold literals. It is also noted that the fact that a two- or one-place operator is not listed here does not preclude its possible importance for some future novel technology or implementation. If a new primitive is considered, new algebraic and logic design techniques may also have to be considered.

2.5.2   Functional Completeness

A fundamental concept is whether a subset of two-place and one-place functions is sufficient to realize all possible p-valued functions. In the common $p = 2$ logic used to design modern digital circuits, there exist a total of only four one-place operators and 16 two-place operators. Due to this limited number of operators, it is relatively easy to define a useful functionally complete algebra. Examples of such algebras are $\langle B, \cdot, +, ', \mathbf{1}, \mathbf{0} \rangle$, $\langle B, \overline{\wedge}\ \mathbf{1} \rangle$, and $\langle B, \overline{\vee}, \mathbf{0} \rangle$. These example algebras are commonly used because they are complete and the operators are efficiently implemented in hardware or software.

In an MVL system with $p > 2$, the number of possible operators and functions increases rapidly. Because of the double exponential increase in the number of possible functions, p^{p^n}, and corresponding increases in the total number of operators, it becomes important to utilize an algebra that is convenient (and generally complete) for the MVL application under consideration. In this section we formally describe the notion of functional completeness and survey some of the more popular algebras that have been utilized in the past.

Definition 2.8 Functional completeness.   A set of functions A over E is termed *functionally complete* if it is possible to define all functions over E as a composition of functions from A.

Definition 2.9 Functionally complete with constants.   A set of functions A that is complete over E only when it is augmented by the constant functions is termed *functionally complete with constants*.

From the discussion in the previous section, $\langle B, \cdot, +, ', \mathbf{0}, \mathbf{1} \rangle$ is functionally complete for $B = \{0, 1\}$. Other well-known complete sets for the two-valued case are those based on the Sheffer Stroke (NAND) $\langle B, \overline{\wedge}\ \mathbf{1} \rangle$, the operator of Peirce (NOR) $\langle B, \overline{\vee}, \mathbf{0} \rangle$, and $\langle B, \cdot, \oplus, ' \rangle$. These former two algebras are more precisely described as being "functionally complete with constants" and the latter as simply "functionally complete". We note that for practical reasons, functionally complete algebras may be specified with the inclusion of redundant constants or operators in their definition. Such specifications are often made for convenience since the additional components may represent corresponding circuit or algorithmic functions that are easily realized in some particular implementation.

2.6   EXAMPLE MVL ALGEBRAS BASED ON NONMODULAR OPERATORS

This section presents a survey of MVL algebras based on logic operations that are not necessarily also modular arithmetic operations over a finite field defined by the logic value set. Many of these algebras are the result of symbolic logicians wishing to define logic systems that provide more expressive reasoning frameworks as compared to traditional binary-valued propositional logic systems. The algebras in this section that were developed more recently were generally motivated by issues in computer system design and analysis.

2.6.1   Lukasiewicz Logic

One of the first published ternary logic systems in a modern form was due to the work of Lukasiewicz in Poland in 1920 [56]. This system was motivated by the consideration of a third logic value of *indeterminate* in addition to the values representing *true* and *false*. In the following description, we will denote the ternary logic values numerically where *false* $\equiv 0$, *indeterminate* $\equiv 1$, and *true* $\equiv 2$. Table 2.6 defines the operations of this logic system.

Originally, Lukasiewicz presented his logic system in terms of only the one-place operator and the $\rightarrow$ (implication) two-place operator defined in Table 2.6. The other three two-place operators $\cdot$, $+$, and $\leftrightarrow$ are easily derived through use of the identities $x + y = (x \rightarrow y) \rightarrow y$, $x \cdot y = \overline{(\overline{x} + \overline{y})}$, and $x \leftrightarrow y = (x \rightarrow y) \cdot (y \rightarrow x)$.

It is important to note that the ternary logic system of Lukasiewicz combined with the operators in Table 2.6 do not form a functionally complete algebra. When the additional one-place operator known as the "T-function" [103,104] is added, then a functionally complete algebra given by $\langle E, \rightarrow, \bar{\ }, T(), \mathbf{0}, \mathbf{1} \rangle$ results where $E = \{0, 1, 2\}$. The T-function is defined in Table 2.7.

Although the original work of Lukasiewicz was based on a $p = 3$-valued logic system, it is possible to generalize the system for the case where $p > 3$ including infinite-valued systems. In such

TABLE 2.6: Operators for ternary Lukasiewicz logic system

x	$\overline{x}$	$\cdot$	0	1	2	$+$	0	1	2	$\rightarrow$	0	1	2	$\leftrightarrow$	0	1	2
0	2	0	0	0	0	0	0	1	2	0	2	2	2	0	2	1	0
1	1	1	0	1	1	1	1	1	2	1	1	2	2	1	1	2	1
2	0	2	0	1	2	2	2	2	2	2	0	1	2	2	0	1	2

TABLE 2.7: Definition of one-place T-function operator

x	$T(x)$
0	1
1	1
2	1

TABLE 2.8: Operators for ternary chained post logic system

x	$C_0(x)$	$C_1(x)$	$C_2(x)$
0	2	0	0
1	0	2	0
2	0	0	2

$\cdot$	0	1	2
0	0	0	0
1	0	1	1
2	0	1	2

$+$	0	1	2
0	0	1	2
1	1	1	2
2	2	2	2

a generalization, it is seen that when $p = 2$, the resulting logic system and algebra is identical to the binary-valued Boolean algebra when $\mathbf{0} \equiv 0$ and $\mathbf{1} \equiv 2$.

2.6.2   Post Logic and Algebra

Although the MVL logic of Lukasiewicz was published one year earlier than the work of Emil Post in 1921 [83], the Post algebras are commonly cited as the first example of a MVL algebra since the "chained Post algebra" is functionally complete and that of Lukasiewicz was not. This set of complete MVL algebras are commonly referred to as "Post algebras". A particular instance is the "chained Post algebra". The chained Post algebra is defined as $\langle E, +, \cdot, \{C_a(x)\}, \mathbf{0}, \mathbf{1} \rangle$ where E is a totally ordered finite set containing p elements $\{0, 1, \ldots, p-1\}$, $\mathbf{0} = 0$ and $\mathbf{1} = p - 1$. The set of one-place operators used in this algebra are the decisive literals, $C_a(x) \forall a \in E$. The chained Post algebra is functionally complete and it is noted that when $p = 2$ (that is, $E = B$), the chained Post algebra is identical to Boolean algebra. Although Boolean algebras may be defined for $p > 2$, they are only functionally complete for $p = 2^n$, $\forall n \in \{1, 2, 3, \ldots\}$, thus the chained Post algebra can be viewed as more general in the sense that it is functionally complete over all MVL systems. One- and two-place logic operations for the ternary Post logic are defined in Table 2.8 where the $\cdot$ operator is commonly referred to as the "min" operator and $+$ the "max" operator.

2.6.3   Bochvar Logic

A variation of the ternary logic system of Lukasiewicz was developed by Bochvar and appeared in his 1939 paper [10]. The variation was due to a difference in the semantical interpretation of the third logic value. The *indeterminate* logic value as defined by Lukasiewicz was construed by Bochvar

TABLE 2.9: Operators for ternary Bochvar logic system

x	$\bar{x}$	$\cdot$	0	1	2	$+$	0	1	2	$\rightarrow$	0	1	2	$\leftrightarrow$	0	1	2
0	2	0	0	1	0	0	0	1	2	0	2	1	2	0	2	1	0
1	1	1	1	1	1	1	1	1	1	1	1	1	1	1	1	1	1
2	0	2	0	1	2	2	2	1	2	2	0	1	2	2	0	1	2

as an *undecidable* value. This resulted in a change in the evaluation of the four two-place operations of ($\cdot$), ($+$), implication ($\rightarrow$), and equivalence ($\leftrightarrow$) as compared to those shown in Table 2.6. Most notably, the two-place multiplicative ($\cdot$) and additive ($+$) operators are no longer the min and max functions, however, when the Bochvarian logic is reduced to a binary-valued logic system by removal of the *undecidable* value, it becomes identical to the binary-valued Boolean logic and is functionally complete. The operators for the ternary Bochvar logic system are given in Table 2.9. As is similar to the Lukasiewicz ternary logic system, additional one-place operators are necessary in order to formulate a functionally complete algebra.

2.6.4 Kleene Logic

Another ternary logic system was proposed by Kleene in 1938 [50]. As is the case of the Bochvarian ternary logic system, the difference in the Kleene system can be considered to be a different interpretation of the third logic value. While the system of Lukasiewicz contained the *indeterminate* value and the system of Bochvar an *undecidable* value, the third value in the Kleene system is interpreted as simply an *unknown* value. This interpretation is analogous to the X value used in the VHDL and Verilog HDLs for initialization of circuit nets. The ternary Kleene logic system may be defined using the operators in Table 2.10.

2.6.5 Allen and Givone Algebra

The MVL algebra introduced by Allen and Givone in 1968 [4] was motivated by the ease of circuit implementation of the corresponding operators in circuits utilizing bipolar junction tran-

TABLE 2.10: Operators for ternary Kleene logic system

x	$\bar{x}$	$\cdot$	0	1	2	$+$	0	1	2	$\rightarrow$	0	1	2	$\leftrightarrow$	0	1	2
0	2	0	0	0	0	0	0	1	2	0	2	2	2	0	2	1	0
1	1	1	0	1	1	1	1	1	2	1	1	1	2	1	1	1	1
2	0	2	0	1	2	2	2	2	2	2	0	1	2	2	0	1	2

sistors (BJTs). Many of the algebras appearing in the literature since the 1960s were developed in order to form a mathematical basis to support MVL circuitry based on a particular implementation technology.

The Allen and Givone algebra is the sextuple $\left\langle E, +, \cdot, x^{[a,b]}, \mathbf{0}, \mathbf{1} \right\rangle$ where E is a finite set containing p elements $\{0, 1, \ldots, p-1\}$, $\mathbf{0} = 0$ and $\mathbf{1} = p-1$. This algebra satisfies the properties of the two-place min and max operators given in Table 2.13.

As for the window literal, it follows from the definition of $+$ and $\cdot$ that

$$x^{[a,b]} = x^{[c,d]} + x^{[e,f]}$$

if, and only if, $a = min(c, e)$, $b = max(d, f)$, $e \leq d+1$ and $c \leq f+1$, and

$$x^{[a,b]} = x^{[c,d]} \cdot x^{[e,f]}$$

if, and only if, $a = max(c, e)$, $b = min(d, f)$, $e \leq d$ and $c \leq f$.

2.6.6   Vranesic, Lee and Smith Algebra

An algebra specified by Vranesic, Lee, and Smith in 1970 [120] can be defined as $\langle E, +, \cdot, x^k, x^{\underline{a}}, \mathbf{0}, \mathbf{1} \rangle$. The one-place operators include the cycle and threshold literal functions. This algebra was motivated based on circuit implementation criteria of ease of minimization algorithms and economical implementation of logic gates that implement the operators. The authors of [120] argue that circuit implementations of threshold functions along with the introduction of a universal clockwise-cycling circuit yield a set of easily implementable logic circuits using technology available in the early 1970s. Additionally, an algorithm was presented for automated optimization of logic covers for functions represented using this algebra and could thus form the basis for efficient synthesis methods.

2.7   EXAMPLE MVL ALGEBRAS BASED ON MODULAR OPERATORS

Modular algebras are those defined with two-place operators that are modular over a finite field of integers where the p truth values of the logic system are the members of a finite field. Several MVL algebras have been developed in the past based on operators that are modular primarily motivated by computing systems design concerns. This section provides a survey of some of these algebras. In terms of binary-valued logic, the most common modular algebra is that often referred to as Reed–Muller logic [78,86] where operators are addition modulo-2 and multiplication modulo-2, which are the Boolean exclusive-OR and AND operators respectively. The binary Reed–Muller algebras can be characterized as the set $\langle B, \oplus, \cdot, \mathbf{1}, \mathbf{0} \rangle$ where $\oplus$ and $\cdot$ represent addition and multiplication modulo-2

TABLE 2.11: Operators for binary Reed–Muller system

$\oplus$	0	1		$\cdot$	0	1
0	0	1		0	0	0
1	1	0		1	0	1

respectively. The unary complementation operation can be conveniently computed as $x' = 1 \oplus x$. Definitions of the Reed–Muller two-place operators are given in Table 2.11.

2.7.1 Cohn Algebra

From a purely algebraic point of view, a modular algebra was developed and reported by Bernstein in 1924 [9]. One of the first functionally complete modular algebras developed for applications to switching functions is described by Cohn in [24]. In this work, classes of canonical forms of switching functions are formulated called Δ-canonical forms. The Δ-canonical forms are expressed in terms of the operations of finite field modular addition and multiplication. The Δ-canonical forms are first formulated in the binary domain and use addition and multiplication operators over the finite field of integers modulo-2 ($\oplus$ and $\cdot$). Later, in Chapter 4 of [24], Δ-forms are generalized to the case of an arbitrary prime modulus. This generalization can be considered to be a case of the binary-valued Reed–Muller algebras extended to an MVL algebraic system where the radix value p is restricted to prime numbers. An example of the Cohn algebra for the ternary case ($p = 3$) leads to the two-place operators defined in Table 2.12.

Because the work of Cohn considered the development of canonical forms of switching functions over rings of integers modulo-p, the resulting matrices and forms were only useful when p is a prime number.

2.7.2 Pradhan Algebra

The algebra defined by Pradhan in 1974 [84] is functionally complete for p-valued logic systems where p is a power of a prime number and can be viewed as a generalization of the Cohn algebras that supported only p values that are prime. The primary contribution of Pradhan was a proof that

TABLE 2.12: Two-place operators for ternary Cohn system

$\oplus$	0	1	2		$\cdot$	0	1	2
0	0	1	2		0	0	0	0
1	1	2	0		1	0	1	2
2	2	0	1		2	0	2	1

TABLE 2.13: Operators for ternary Dubrova and Muzio logic system

x	$C_0(x)$	$C_1(x)$	$C_2(x)$
0	2	0	0
1	0	2	0
2	0	0	2

$\cdot$	0	1	2
0	0	0	0
1	0	1	1
2	0	1	2

$\oplus$	0	1	2
0	0	1	2
1	1	2	0
2	2	0	1

modular algebras for any modulus $N = p^m$ forms a functionally complete set. This proof is based on concepts in coding theory.

2.7.3   Dubrova and Muzio Algebra

In [27], a modular algebra is defined for a modulus p for any value of p that is a positive integer. The algebra is based on the operations of modular-addition, the two-place minimum function, and a set of unary (literal) operators. Notationally, this algebra is defined as $\langle E, \oplus, \cdot, \{C_a(x)\}, \mathbf{0}, \mathbf{1} \rangle$ where E is a totally ordered finite set containing p elements $\{0, 1, \ldots, p-1\}$, $\mathbf{0} = 0$ and $\mathbf{1} = p-1$. The set of one-place operators used in this algebra are the decisive literals, $C_a(x) \forall a \in E$. It is noted that this algebra is not purely modular since it only utilizes modular addition and multiplication (denoted by $\cdot$) that is the nonmodular product used in the chained Post algebra. This algebra also includes the set of unary operators sometimes referred to as "Post literals," $C_a(x)$. An example of the Dubrova and Muzio algebra for the ternary case ($p = 3$) leads to the operators defined in Table 2.13.

The Dubrova and Muzio algebra has the advantage that implementation of the non-modular product operator is implementable in CMOS current-mode based logic circuits with only five transistors independent of the value of p whereas a modular product operation requires 16 transistors for $p = 3$.

2.8   SUMMARY OF MVL ALGEBRAS

Many different MVL logic systems and associated algebras are possible due to the exponentially increasing number of operators with respect to the cardinality of the MVL logic value set, p. The algebras surveyed in this chapter were generally developed for a particular application domain. Many of the more recently developed algebras were developed for the purpose of modeling MVL circuits based upon particular electronic components to be used as primitive circuit elements in their construction. The earlier algebraic systems were developed by symbolic logicians with the goal of specifying a more descriptive logic system for reasoning. As new circuit devices or other application domains emerge requiring the use of discrete logic values for $p > 2$, it is likely that some of these algebraic systems, or perhaps new ones will be the most useful; however, the central concept of functional completeness is still the underlying concept to be considered.

CHAPTER 3

Functional Representations

3.1 LOGIC TABLES

The use of tables to represent functions and connectives is well known to those familiar with binary logic design.

Example 3.1. Table 3.1 represents the sum (S) and carry (C) of three bits (a,b, and c).

The tabular enumeration of a function is, at least in theory, possible provided the function has finite domain and range. Hence, the use of tables extends directly to the multiple-valued logic (MVL) case as shown in Table 3.2, which represents the mod-3 sum S of two ternary digits, a and b. The tabular representation of a p-valued, n-input, m-output function has p^n rows and $n + m$ columns. A common use of tables is in representing logic connectives (operators) as in Chapter 2.

The table approach clearly becomes impractical, at least for hand use, as p or n increase. For example, the table representing a function with 5 four-valued input variables has 1024 rows. Despite this growth in the tabular representation, there are situations where it is useful in programming. It should be noted that it is not necessary to store the input patterns and it is possible to store a function as a simple output vector. For example, the function represented in Table 3.2 has vector representation $(0, 1, 2, 1, 2, 0, 2, 0, 1)'$. Each value can be represented using $\lceil \log_2 p \rceil$ bits so a p-valued, n-input,

TABLE 3.1: Representation of the sum and carry of three bits

a	b	c	C	S
0	0	0	0	0
0	0	1	0	1
0	1	0	0	1
0	1	1	1	0
1	0	0	0	1
1	0	1	1	0
1	1	0	1	0
1	1	1	1	1

TABLE 3.2: Representation of the mod-3 sum of two ternary values		
a	b	S
0	0	0
0	1	1
0	2	2
1	0	1
1	1	2
1	2	0
2	0	2
2	1	0
2	2	1

m-output function can be represented using $m \times p^n \times \lceil \log_2 p \rceil$ bits of storage. This of course comes at some computational cost for packing and unpacking the representation.

3.2   HYPERCUBES

An alternative to simple enumerative tables is to consider the input patterns (minterms) as points in an n-dimensional space as shown by the binary examples in Fig. 3.1. A function is represented by marking each minterm as true or false. These hypercube constructions show the interrelations between the input patterns and can thus show particular relationships. For example, product terms

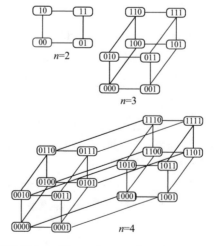

FIGURE 3.1: Binary hypercubes for $n = 2, 3, 4$.

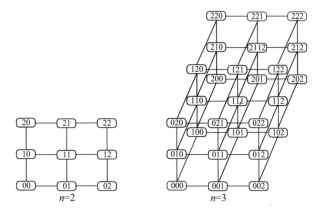

FIGURE 3.2: Ternary hypercubes for $n = 2, 3$.

are identified as sets of adjacent points on the hypercube. Clearly, such a set must contain a power of 2 points.

More complex properties are also representable on the hypercube. For example a function is linearly separable if it is possible to draw an $n - 1$ dimensional plane through the n-dimensional hypercube such that all minterms on one side of the plane are false while all minterms on the other side of the plane are true.

The hypercube concept readily extends to higher radix functions as shown in Figure 3.2 for the ternary case for $n = 2, 3$. In the p-valued case, each point in the hypercube is assigned a value $0, \ldots, p - 1$ in order to define the function.

Clearly, hypercube representations become hard, if not prohibitive, to draw and interpret as n and p increase. Their use is generally restricted to small illustrative examples and they are not used for practical logic design, even for small problems.

3.3   MAPS

A function map is a projection of the appropriate hypercube onto the plane. As such it is both easier to draw and to interpret and can be used quite effectively, at least for small examples.

3.3.1   Binary Maps

The reader familiar with binary logic design will be familiar with the concept of a Karnaugh map (K-map) [47]. K-maps can be drawn in various ways. Figure 3.3 shows a common way for binary K-maps for functions of 2, 3 and 4 variables. K-maps are labeled so that horizontally and vertically adjacent cells have labels that differ in a single bit position. The edges of the map wrap around from left to right and from top to bottom with respect to this adjacency property. With practice, K-maps can be used effectively for functions with up to six variables.

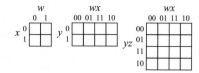

FIGURE 3.3: Binary K-maps for $n = 2, 3, 4$.

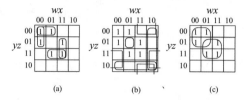

(a)   (b)   (c)

FIGURE 3.4: Three normal form solutions: (a) sum-of-products, (b) product-of-sums and (c) XOR sum-of-products.

It is straightforward to identify a sum-of-products (SOPs) (disjunctive normal) form for a function by identifying blocks of adjacent minterms (1's) on a map where each block includes a power of 2 1's and every 1 in the function is covered at least once.

Example 3.2. The blocks identified in Fig. 3.4(a) identify the SOPs form $w'y'z' + w'x'y' + xyz + wxz$. Blank squares in Fig. 3.4 denote logic 0.

Circling 1's on a map is in fact the repetitive application of the identity $\bar{x}\alpha + x\alpha = \alpha$ where α is a product term not involving variable x. A product-of-sums (POSs) (conjunctive normal) form is found by considering the maxterms (0's) rather than the minterms and applying the dual identity $(\bar{x} + \alpha)(x + \alpha) = \alpha$. Note that in this case, the variables are inverted when writing the terms for the blocks.

Example 3.3. The blocks in Fig. 3.4(b) thus identify the expression $(w' + z)(w' + x)(x + y')(y' + z)(w + x' + y + z')$.

It is less well known that K-maps can also be used to find XOR sum-of-products solutions. In this case, each 1 must be covered an odd number of times and each 0 must be covered zero or an even number of times.

Example 3.4. The solution indicated in Fig. 3.4(c) is $w'y' \oplus xz$. This result is simpler than either the sum-of-products or product-of-sums solution with respect to number of terms and number of literals.

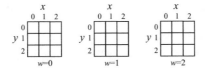

FIGURE 3.5: Ternary map for $n = 2$.

FIGURE 3.6: Ternary map for $n = 3$.

3.3.2 Multiple-Valued Maps

As illustrated above, the key feature of binary K-maps is that only one variable changes when moving from a cell to an adjacent cell (ignoring diagonal adjacencies). The ternary map in Fig. 3.5 has a similar property. How best to extend the ternary map to three variables is not obvious. One possibility is shown in Fig. 3.6. Using such a map is not straightforward and is akin to using a five or six variable map for the binary case. Other possibilities have been explored, see for example Hurst [43], but are no more satisfactory.

Systematic procedures equivalent to those used to find sum-of-products or product-of-sums forms can be applied to multiple-valued maps for small problems and these approaches give insight into developing CAD programs for handling larger problems.

Extending these ideas to more variable or to the case of $p = 4$ or higher is even more problematic. One concludes that except for small ternary examples and for illustrative purposes, multiple-valued maps are not particularly helpful.

3.4 CUBE NOTATION

Roth [89] considered algebraic topological methods for the simplification of binary switching functions developing a positional notation and a set of operators that are directly usable in CAD programs. In addition, the ideas are readily extended to and remain very practical for multiple-valued logic. This approach is referred to as *cubical calculus* or, somewhat less formally as *cube notation*. The presentation here is informal and is meant to provide context for the later discussion of decision diagrams (DD). Readers interested in a more formal development should consult the literature.

3.4.1 Binary Cubes

Consider a binary function $f(x_1, \ldots, x_n)$. A *product term* is a conjunction where each x_i appears negated, affirmed, or does not appear. There are clearly 3^n product terms for n variables.

Definition 3.1. An n-variable product term can be uniquely represented by a **cube** which is an ordered set of n symbols with 0, 1, $-$ denoting that the corresponding variable is negated, affirmed or does not appear respectively.

Example 3.5. Given $n = 3$, the product term $x_1' x_3$ is represented by the cube 0–1. Note that a cube for n variables always contains n symbols.

Definition 3.2. A cube **covers** the set of minterms found by substituting 0's and 1's for dashes in all possible ways. A cube with k dashes covers 2^k minterms, hence a cube with no dashes covers a single minterm.

Definition 3.3. Two cubes a and b are **opposed** in position i if one of a_i and b_i equals 1 and the other equals 0.

Definition 3.4. Two cubes are **adjacent** if they are opposed in exactly one position.

The term *cube* arises naturally from the fact a cube with k dashes covers 2^k minterms which form a k-dimensional subcube within the n-dimensional hypercube. The term *adjacent* comes from the positioning of the two cubes in the appropriate hypercube.

Definition 3.5. The **cubical complex** of a function is the set of all cubes that cover only minterms for which the function is 1.

Since a map is a projection of the hypercube, a cube represents a block of adjacent 1's on the map which as noted above must contain a power of 2 minterms. The cubical complex thus includes the set of all possible blocks of 1's including the subblocks of each block.

Definition 3.6. A cube c is **prime** if no other single cube in the cubical complex for the function covers all the minterms that are covered by c. A prime cube corresponds to the better known term **prime implicant**.

Definition 3.7. A **cover** of a cubical complex is a subset C of the cubes in that complex such that every minterm in the complex is covered by at least one cube in C. C is a **prime cover** if every cube in C is prime. C is a **minimal cover** if the number of cubes in C is minimal with respect to all covers of the complex. It is readily shown that there must be a minimal cover composed of only prime cubes.

Note that the above can be alternatively defined in terms of the maxterms and sum terms of the function.

3.4.2   Multiple-Valued Cubes

To see how to extend cube notation to the multiple-valued case, it is useful to view binary cubes in a somewhat different way from the traditional view outlined above. A dash in a binary cube denotes the fact that the corresponding variable can take the value 0 or 1, whereas the symbols 0 and 1 denote the fact the variable takes that value. The symbols in the cube can be viewed as sets.

Example 3.6.   Consider the cube,

$$00 - 1 - 0 - 1$$

Viewed as sets, the cube is

$$\{0\}\{0\}\{01\}\{1\}\{01\}\{0\}\{01\}\{1\}$$

Adopting the more compact notation that a set of one element is written without braces, gives

$$00\{01\}1\{01\}0\{01\}1$$

This set notation is easily extended to the multiple-valued case as shown in the following definition:

Definition 3.8.   An n-variable **multiple-valued cube (MV-cube)** is an ordered set of sets $c = \{c_1, c_2, \ldots, c_n\}$ where each c_i contains the set of values the ith input variable can assume.

Example 3.7.   Consider Fig. 3.7. Suppose that for the four minterms highlighted on the hypercube, the function is to assume the value 2 as shown on the map. The corresponding cube is $\{02\}2\{01\}$. The fact that the function is to assume the value 2 is not captured and requires a further extension to the binary notation.

The notion of opposing binary cubes extends to MV-cubes as follows:

Definition 3.9.   Two MV-cubes a and b are **opposed** in position i if $a_i \cap b_i = \phi$.

A multiple-valued function can be viewed as having a set of cubical complexes, with one complex for each output value. Alternatively, it can be considered to have a single complex where each cube is augmented with the appropriate output value. We adopt this second approach. Note that the cubes for output value 0 are usually omitted, although the function remains fully specified if the cubes for any single value are omitted.

The definitions above of *covering*, *adjacency* and *prime cube* for binary cubes extend to multiple-valued cubes in the obvious ways. Likewise the definitions of *cover*, *prime cover* and *minimal cover* apply to the cubical complex of a multiple-valued function.

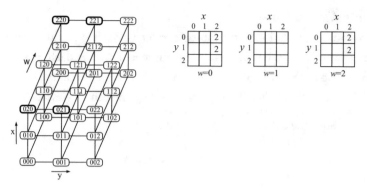

FIGURE 3.7: Example mV-cube.

3.4.3   Cube Representation of Functions

3.4.3.1 Totally specified functions. A totally specified binary function $f(x_1, \ldots, x_n)$ can be defined by any cover C of the cubical complex of f. Note that this includes defining the function by listing its minterms. The classical binary minimization problem is to find a minimum prime cover. This corresponds to finding a minimal sum-of-products (disjunctive normal form) expression.

There are other binary minimization problems including product-of-sums, which is the dual of the sum-of-products problem, and XOR sum-of-products minimization. In the latter case a minimal size subset of the cubical complex must be found such that every minterm for which $f(x_1, \ldots, x_n) = 1$ is covered an odd number of times and every minterm for which $f(x_1, \ldots, x_n) = 0$ is covered zero or an even number of times.

The situation is more involved for multiple-valued functions since it turns out the cubes included in the cubical complex depend on the operator set being considered. For example, suppose we are interested in finding a minimal MAX-of-MIN expression for a function (see Tables 2.2 for the definition of MAX and MIN). Since the operation combining the output values associated with cubes is MAX, the minterms for value k act as don't-cares for cubes with output values j for all $j < k$. This means those minterms can be used when possible to form larger cubes using less variables. This don't-care property does not hold for all forms of MVL expressions.

3.4.3.2 Partially specified function. A partially specified function is one where the output value associated with one or more minterms is a don't-care condition. The notion of a don't-care in the binary case is quite straightforward since it simply means the function can take either the value 0 or 1. For minimization problems, the cubical complex is expanded to include cubes involving 1 and don't-care minterms where each such cube includes at least one 1-minterm. The subset of the complex is then chosen to have a minimum number of cubes with the appropriate criterion applied with regard to the 1-minterms only.

In the multiple-valued case, the notion of a don't-care output is much broader. It can mean the function can take any value $0, \ldots, p-1$, which is equivalent to the binary case, or it can mean that the function can take one of a proper, nonempty subset of $\{0, \ldots, p-1\}$, which has no binary counterpart. The former has been termed a *full* don't-care, whereas the latter has been termed a *partial* don't-care. A representation capturing this full flexibility is introduced below.

3.4.3.3 Multiple-output functions. Practical design most frequently involves situations where there are a number of outputs dependent on a common set of inputs. We here consider multiple-output, multiple-valued functions recognizing that this includes multiple-output binary functions as a special case. It also includes the often considered situation of multiple-output functions with multiple-valued inputs and binary outputs.

Again, set notation proves helpful. A multiple-output multiple-valued function can be fully defined by a list of MV-cubes where each cube is augmented by sets which identify the values each of outputs can take for the minterms covered by the cube. This approach is formalized in the following definition:

Definition 3.10.   An **augmented MV-cube** c for n-inputs and m-outputs is a pair of ordered sets of sets c^{in} and c^{out} with $c^{in} = \{c_1^{in}, c_2^{in}, \ldots, c_n^{in}\}$ where each c_i^{in} contains the set of values the ith input variable can assume, and $c^{out} = \{c_1^{out}, c_2^{out}, \ldots, c_m^{out}\}$ where c_j^{out} is the set of values the jth output can assume.

3.4.4   Operations on Cubes

We consider operations on MV-cubes and augmented MV-cubes which includes operations on binary cubes.

The intersection of two MV-cubes a and b is a cube c covering the minterms common to a and b. It is given by:

$$c = \{a_1 \cap b_1, a_2 \cap b_2, \ldots a_n \cap b_n\}, \; if \; no \; a_i \cap b_i = \phi,$$

$$\phi, \; otherwise.$$

For augmented cubes, intersection is extended as follows:

$$c^{in} = \{a_1^{in} \cap b_1^{in}, a_2^{in} \cap b_2^{in}, \ldots, a_n^{in} \cap b_n^{in}\}, \; if \; no \; a_i^{in} \cap b_i^{in} = \phi,$$

$$\phi, \; otherwise.$$

$$c^{out} = \{a_1^{out} \cup b_1^{out}, a_2^{out} \cup b_2^{out}, \ldots, a_n^{out} \cup b_n^{out}\}, \; if \; a^{in} \cap b^{in} \neq \phi,$$

$$\phi, \; otherwise.$$

The intersection of cubes can be used to identify a cube a is subsumed by a second cube b if the intersection of the two cubes equals a.

The binary $a\#b$ extraction operation defined by Roth [90] yields a list of cubes covering all minterms covered by a that are not covered by b. It is extended to MV-cubes as follows:

- $a\#b = a$ if $a_i \cap b_i = \phi$ for any i,
- $a\#b = \phi$ if $a_i = b_i$ for all i,
- $a\#b$ is the list of cubes including $\{a_i, a_2, \ldots, a_i - (a_i \cap b_i), \ldots, a_n\}$ for all i such that $a_i - (a_i \cap b_i) \neq \phi$.

Given a list of MV-cubes $A = \{A_1, A_2, \ldots, A_t\}$, $A\#b = (A_1\#b), (A_2\#b), \ldots, (A_t\#b)$ where ',' denotes list concatenation. If A has no subsuming cubes neither does $A\#b$.

Given a list of MV-cubes $B = \{B_1, B_2, \ldots, B_s\}$, $A\#B = (\cdots((A\#B_1)\#B_2)\cdots\#B_s)$. Again if A has no subsuming cubes neither does $A\#B$.

Two cubes are disjoint if the sets of minterms they cover are disjoint. A cube list is disjoint if its members are mutually disjoint. An important application of the # operation is to covert a cube list to a disjoint cube list.

3.4.4.1 Minimal covers.

As noted above, finding minimal cube covers is equivalent to function minimization which is known to be a very complex problem. We here present only some basic ideas as illustrations of the use of cubes. Only totally specified single-output functions are considered. The interested reader can consider the extension to the multiple-output and partially specified cases using the concepts outlined above.

Consider a cover C_i of a cubical complex. Construct a new cover C_{i+1} by considering all possible pairs of cubes a and b from C_i and putting the new cube $\{a_1, a_2, \ldots, a_k \cup b_k, \ldots, a_n\}$ into C_{i+1} if a and b are adjacent with respect to the kth variable. A cube from c_i is copied to C_{i+1} if it does not combine with any other cube from C_i. If C_0 is initially set to a complex of minterms, and the above is iterated until the new cover is identical to the prior cover, that cover contains all the prime cubes. For binary functions, C_0 contains the minterms for which the function is 1. For multiple-valued functions, C_0 contains all the minterms for which the function is not 0, and cubes are only combined if they have the same output value.

This approach is the basic Quine–McCluskey procedure [65,97] and can be seen to be an exhaustive enumeration approach. For multiple-valued functions, it treats the output values as separate covering problems and does not use the minterm don't-care property noted above.

This approach can be modified to an iterated consensus [57] type approach by changing the rule for forming a new cube to $\{a_1 \cap b_1, a_2 \cap b_2, \ldots, a_k \cup b_k, \ldots, a_n \cap b_n\}$ with the continued requirement that the cubes be adjacent with respect to the kth variable. In this case, C_0 can be any cover of the complex. The cubes from C_i are copied to $C_{i+!}$ unless they are subsumed by another cube in C_{i+1}. Again, the final cover is the list of prime cubes.

Both of the above methods yield all prime cubes. A minimal cover must be found using a minimal set covering technique. Roth [90] has presented an alternative approach using the # operation.

Modern minimization procedures are much more complex than the simple ideas just described and are effective for quite large problems. However, certain issues remain difficult. For example, the cubes in a minimal cover are not ordered and the minimal cover is not unique. This makes it difficult to determine if two functions represented by cube covers, minimal or otherwise, are equivalent. We will see in Section 3.5 how this is handled by DDs.

3.4.5   Function Operations

It is often necessary to perform logical operations on a function or on a set of functions. This can be used, for example, to determine the function performed by a given circuit.

Clearly, given a function represented by a list of MV-cubes, a unary operator can be applied to an input or an output by applying the appropriate operation to the set corresponding to the input or output in each augmented MV-cube in the specification.

Suppose we wish to evaluate $h = f \; op \; g$ where f is represented by a cube list F, g is represented by a cube list G, and op is some logical operation. The goal is to find a cube list H representing h. This can be accomplished by considering every F_i and G_j pair of cubes. If the input sets of F_i and G_j intersect, a cube H_k is created whose input set is that intersection. The output set for H_k must be formed one output at a time by combining the appropriate sets from F_i and G_j according to the nature of the op. If op is MIN for example, each output set for H_k consists of the values from $F_i \cup G_j$ up to the minimum of the maximum values for those two sets. For MAX, the approach is the same but the new set contains the values from the two given sets down to the maximum of the minimum values for those sets.

Once again, the nonuniqueness of the cube-based representations limits this approach due to the difficulty in determining equality of functions. For example, we can use the above approach to find cube list descriptions for two circuits, but it remains a difficult problem to determine if those two lists are equivalent and that the two circuits realize the same function.

3.5   DECISION DIAGRAMS

DDs are a state-of-the-art representation for binary logic functions. This section will review the basic concepts of binary decision diagrams (BDD) and show how they are extended to the multiple-valued case. Techniques to achieve efficiency in the construction and manipulation of multiple-valued decision diagrams (MDD) are discussed. MDD also form a basis for quantum multiple-valued decision diagrams (QMDD) presented in Chapter 5.

3.5.1   Binary Decision Diagrams

BDD are a graphical representation for binary functions, sets, and relations. They can also be used in the representation and manipulation of vectors and matrices.

3.5.1.1 The basic structure. The basic BDD concept was introduced by Lee [53] and further studied and refined by Akers [2]. The full potential of BDD for representation and computation was identified in the seminal work of Bryant [13] which introduced two key concepts:

- *Ordered:* There is a global ordering of the variables such that the variables on every path from the start vertex to a terminal vertex adhere to that ordering and no variable appears more than once on any single path.

- *Reduced:* Common subgraphs are shared and redundant subgraphs are removed.

BDD satisfying these two rules are called reduced ordered binary decision diagrams (ROBDD). Bryant showed the ROBDD representation of a given function is unique up to variable ordering. This unique representation property is critical to the power of ROBDD. For example, if one builds ROBDD for two combinational circuits using the same variable ordering, they will be the same if, and only if, the two circuits implement the same function. Further, as will be shown later, using what are now standard implementation techniques, it is possible to tell if two ROBDDs are identical by checking only the edges pointing to the start (top) vertices of the graphs, comparison of the full structures not being required.

Recall the Shannon decomposition of a Boolean expression P given by $P = x'P_0 + xP_1$ where x is a variable in P which does not appear in P_0 or P_1 (see Theorem 2.1). A BDD is created by repeated application of Shannon decompositions.

Example 3.8.   Consider the Boolean function given by $x_0x_1 + x_0x_2 + x_1x_2$ which the reader will recognize as the 3-bit majority function. Repeated application of Shannon decompositions over x_2 then x_1 and then x_0 can be captured as a tree as shown in Fig. 3.8(a) where each nonterminal vertex is labeled by the decision variable for the Shannon decomposition and has two outgoing edges corresponding to that variable assuming the value 0 (dashed) and the value 1 (solid). Two redundant subgraphs are outlined by solid rectangles and two identical subgraphs are outlined by dashed rectangles in Fig. 3.8(a). Taking these observations into account the tree can be redrawn as the ROBDD shown in Fig. 3.8(b). It can be seen that a ROBDD is a directed acyclic graph (DAG). Throughout the rest of this chapter we will assume all decision diagrams are reduced and ordered.

3.5.1.2 Shared BDD. In many applications, several functions must be represented and manipulated simultaneously. BDD offer a significant advantage in this regard, since as long as the functions adhere to a single variable ordering, common subfunctions (subgraphs) can be shared. Conceptually,

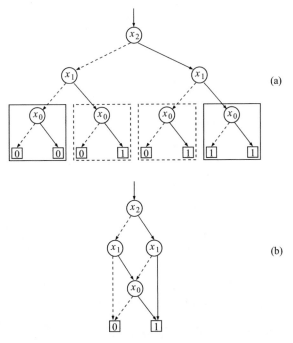

FIGURE 3.8: (a) Binary tree and (b) ROBDD for 3-bit majority function.

a shared BDD is a single DAG structure with an incoming edge to identify the start vertex for each function. If two functions happen to be identical, the start edges point to the same vertex, and in fact that equality is verified by that simple fact without having to traverse the BDD for each function. This makes shared BDD highly efficient for verification purposes.

3.5.1.3 Variable ordering. The variable ordering chosen can affect the size of a BDD.

Example 3.9. Figure 3.9 shows the BDD for $f(x_0, x_1, x_2, x_3) = x_0 x_1 + x_2 x_3$ for two variable orderings $x_0 \prec x_1 \prec x_2 \prec x_3$ and $x_0 \prec x_2 \prec x_1 \prec x_3$. The situation grows worse as more variables are added, *e.g.*, $f(x_0, x_1, x_2, x_3, x_4, x_5) = x_0 x_1 + x_2 x_3 + x_4 x_5$ for the orderings $x_0 \prec x_1 \prec x_2 \prec x_3 \prec x_4 \prec x_5$ and $x_0 \prec x_2 \prec x_4 \prec x_1 \prec x_3 \prec x_5$.

Example 3.10. A more compelling example is the growth in the BDD for the n sum functions and the carry function for the addition of two n-bit numbers $a_{n-1}, \ldots, a_1, a_0$ and $b_{n-1}, \ldots, b_1, b_0$. If the variables are ordered $a_0 \prec a_1 \prec \cdots \prec a_{n-1} \prec b_0 \prec b_1 \prec \cdots \prec b_{n-1}$, the size of the BDD grows exponentially as n increases. For the ordering $a_0 \prec b_0 \prec a_1 \prec b_1 \prec \cdots \prec a_{n-1} \prec b_{n-1}$, the size of the BDD grows linearly as n increases. This is because the second ordering places symmetric variable pairs together in the ordering. The same effect is present for the example in Fig. 3.8. In fact, empirical

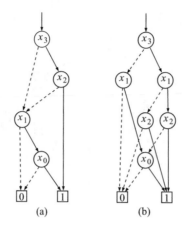

FIGURE 3.9: BDD for $f(x_0, x_1, x_2, x_3) = x_0 x_1 + x_2 x_3$ for variable orderings (a) $x_0 \prec x_1 \prec x_2 \prec x_3$ and (b) $x_0 \prec x_2 \prec x_1 \prec x_3$.

evidence has shown that grouping symmetric variable or almost symmetric variables together leads in general to smaller BDD. There is no proof that this is always the case.

There are $n!$ orderings of n variables. Of course, some are equivalent if there are symmetric variables. It is thus not surprising that, except for small problems, it is computationally infeasible to find a variable ordering guaranteed to minimize the number of vertices in a BDD.

A number of heuristic solutions to the BDD variable ordering problem have been proposed. An efficient approach known as *sifting* was presented by Rudell [92]. Sifting is particularly efficient and effective, and has become the standard variable ordering method for DDs. It has been extended, for example to include reordering of variables by symmetric groups, and is used in many BDD software implementations including the widely used Colorado University decision diagram (CUDD) package developed by Somenzi [106].

Variable reordering, sifting in particular, is often applied dynamically. In this mode, the BDD are build according to a given order, often simply in lexicographic order of the variable names. If the number of vertices reaches a set limit, the BDD package automatically applies a variable reordering strategy hoping to reduce the vertex count for the BDD. Vertices not needed for the new ordering are recycled.

Variable reordering must be applied across a shared BDD but this does not complicate matters and is in fact what is wanted to preserve the uniqueness of the representation across the functions.

A more detailed discussion of variable ordering, sifting in particular, will be presented below once we have discussed the extension of BDD techniques to the multiple-valued case.

3.5.1.4 Edge complements. A simple yet very powerful addition to the BDD structure is to add complement operations to certain edges [77]. The result is that only one of a function and its

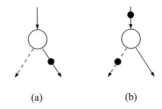

(a)   (b)

FIGURE 3.10: Illustration of a transformation (a) to (b) required for complemented edges.

complement needs to be stored. This means that not only common subfunctions can be shared as shown above, but so can a subfunction and the complement of that subfunction. This can significantly reduce the size of the representation for some functions.

Various rules can be used to apply complemented edges to BDD. For example, the CUDD package [106] uses the following rules:

- There is a single terminal vertex with value 1.

- A 1-edge from a nonterminal vertex cannot be complemented.

- The edge pointing to the start vertex for a BDD may have an edge complement.

An edge complement means the edge points to the complement of the function represented by the subgraph the edge points to. As a result if one follows a path through the BDD corresponding to a particular assignment to the variables, the final function value is 1 if there is an even number of complemented edges on the path, and 0 if there is an odd number. An edge complement will be indicated as a black circle on the edge.

The rule that states there are no complements on 1-edges requires some adjustments as a BDD is built. For example, the situation in Fig. 3.10(a) must be transformed to the structure depicted in Fig. 3.10(b). There are three similar situations to consider as the reader can readily verify. Consider the BDD in Fig. 3.11(a) which the reader can verify represents $f(x_0, x_1, x_2) = x_0 \oplus x_1 \oplus x_2$. Using complemented edges, the same function is realized as shown in Fig. 3.11(b). It is readily shown that for any given function the BDD using complemented edges applied using the rules described above is unique up to variable reordering. Alternative rules, for example a single terminal vertex labeled 0 and no complements on 0-edges also work and again yield unique representations. The unique representations depend on the rules used and also on whether complemented edges are used at all. The important thing is to choose an approach and apply it consistently to ensure the representation of any given function is unique. The sifting technique for variable reordering is directly applicable to BDD with complemented edges.

3.5.1.5 The if-then-else operator. BDDs are frequently constructed using one or a small set of primitive operations. For example, many packages construct diagrams by implementing the normal logic

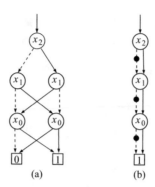

FIGURE 3.11: BDD for $x_0 \oplus x_1 \oplus x_2$ without (a) and with (b) complemented edges.

operations using the single if-then-else (*ITE*) primitive defined as

$$ITE(a, b, c) = if \ a \ then \ b \ else \ c$$

For example, $NOT(a) = ITE(a, 1, 0)$; $AND(a, b) = ITE(a, b, 0)$; and $OR(a, b) = ITE(a, 1, b)$. Employing two *ITE* we have $XOR(a, b) = ITE(a, ITE(b, 1, 0), b)$. More detail is given below where we discuss the CASE operator which is the multiple-valued extension of ITE.

3.5.2   Multiple-Valued Decision Diagrams

The extension of BDD to the multiple-valued case is conceptually straightforward and results in a multiple-valued decision diagram (MDD). The principal difference is that each vertex has p outgoing edges rather than 2 as in the binary case.

   We here describe the basic structure and then outline the key implementation issues and strategies. The purpose is to provide the reader insight into MDDs and their implementation and not to describe a particular implementation in full detail.

3.5.2.1 The basic structure.   Consider a p-valued totally specified function $f(x_0, x_1, \ldots, x_{n-1})$ where the x_i are p-valued. The set of values is $\{0, \ldots, p - 1\}$. Such a function can be represented by an MDD with p terminal vertices each labeled by a distinct logic value. Every nonterminal vertex is labeled by an input variable and has p outgoing edges, one corresponding to each logic value. Note that we here for simplicity only consider homogeneous MDD where all variables and functions are p-valued. The reader can readily see how the techniques apply to the nonhomogeneous case.

   As for BDD, an MDD is ordered if the variables adhere to a single global ordering on every path from the start vertex to a terminal vertex, and no variable appears more than once on any path. Further, a reduced MDD has no vertex where all p outgoing edges point to the same vertex and no duplicated subgraphs. Clearly, no duplicated subgraphs exist if, and only if, no two nonterminal

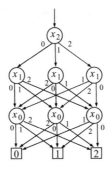

FIGURE 3.12: MDD for three-valued, three-variable sum.

vertices labeled by the same variable, have the same direct descendants. With proper management, reduction can be achieved as the MDD is built. We assume all MDD are reduced and ordered as that is the case of practical interest. A shared MDD is the direct extension of the shared BDD concept with common subgraphs (subfunctions) shared amongst multiple functions.

Example 3.11. As an example of an MDD, consider the sum mod-3 of three three-valued variables. The MDD is shown in Fig. 3.12. It is instructive to compare the structure of this MDD to the BDD in Fig. 3.11(a) which represents a comparable binary function. The use of edge operations to reduce the complexity of the MDD in a manner analogous to that used for the BDD is discussed later.

It has been shown [72], that $R(n, p)$, the maximum number of nonterminal vertices in a p-valued, n-variable MDD is given by

$$R(p, n) = \frac{p^{n-k} - 1}{p - 1} + p^{p^k} - p$$

where

$$k = \lfloor \log_p (n - \lfloor \log_p n \rfloor) \rfloor$$

For $p = 2$ this is equivalent to a separately developed result for BDD [39]. This is a theoretical bound, which while proven constructively, i.e., there exist functions that reach the bound, it is not known how functions of practical interest in general behave in terms of BDD or MDD size.

3.5.3 MDD Implementation Techniques

Many binary techniques, or extensions thereof, are useful when implementing a package for the creation and manipulation of MDD. In this section, we consider implementation strategies for MDD including additions and refinements to the very basic structure described above.

3.5.3.1 Basic data structure. The type of structure required for MDD vertices (expressed in C) is:

```
typedef struct vertex *DDedge;
typedef struct vertex *DDlink;
typedef struct vertex {
   char value;
   DDlink next;
   DDedge edge[0];
}vertex;
```

where the components are as follows:

- *Value:* This component is the index of the variable labeling a nonterminal vertex or the value associated with a terminal vertex.

- *Next:* This pointer is used to chain a vertex into linked lists for memory management and unique table management.

- *Edge:* This is an array of DDedges which is declared empty but which is actually allocated to have p edges.

Note that DDedge and DDlink are similar pointer types but are used in different contexts – edges for the MDD structure and links for vertex management.

The vertex structure shown is very similar to the structure used in binary decision diagram packages. The principal difference is the specification of an array of edges rather than a fixed number (2 in binary). Since the dimension of the array is assigned dynamically when a vertex is created, the array must be at the end of the structure.

3.5.3.2 The CASE operator. A DD is typically built from a functional, algebraic or circuit specification. This requires procedures to apply logic operations on two or more DD. As mentioned above, the ITE operation is often used for this purpose for the binary case as it can be used to implement the standard binary operations: NOT, AND, OR, XOR *etc.*

The MVL generalization of ITE is $CASE(x, y_0, y_1, y_2, \ldots, y_{p-1}) = y_x$, the implementation of which is outlined in Algorithm 3.2 where the following notation is used:

- p is the radix of the logic system being considered, so every nonterminal vertex has p outgoing edges;

- $v(e)$ denotes the vertex edge e points to;

- $x(e)$ denotes the variable that labels the vertex which edge e points to ($x(e)$ returns a result that compares as earlier in the order than any variable if e points to the terminal vertex);

- $E_i(e)$ denotes the ith edge out of the vertex that e points to;
- $term(e)$ denotes a Boolean test that is true if edge e points to the terminal vertex;
- if e points to a terminal vertex, $value(e)$ denotes the value of that vertex; and
- a and $b_0, \ldots, b_{p-1}$ denote the top edges of the MDD to be combined using the $CASE$ function.

Algorithm 3.1 Implementation of MVL CASE.

$CASE(a, b_0, b_1, \ldots, b_{p-1})$
  if$(term(v(a)))$ return$(b_{value(a)})$
  $top =$ highest of $x(a), x(b_0), \ldots, x(b_{p-1})$ in the variable order
  for $0 \le i \le p - 1$
    if$(v(a) = top)w = E_i(a)$
    else $w = a$
    for $0 \le j \le p$
      if$(v(b_j) = top)y_j = E_i(b_j)$
      else $y_j = b_j$
    $z_i = CASE(w, y_0, \ldots, y_{p-1})$
  return a vertex with selection variable top and
    outgoing edges $z_0, \ldots, z_{p-1}$

Common MVL logical operations can be expressed in terms of CASE. For example, for $p = 4$,

$$MIN(a, b) = CASE(a, 0, CASE(b, 0, 1, 1, 1), CASE(b, 0, 1, 2, 2), b)$$

$$MAX(a, b) = CASE(a, b, CASE(b, 1, 1, 2, 3), CASE(b, 2, 2, 2, 3), 3)$$

Other common logic operations can be implemented in a similar fashion.

3.5.3.3 Edge operators. The idea of edge complements for BDD is readily extended to MDD. In theory, an edge operation can be any unary operation, but it is of course necessary to ensure that a set of normalization rules can be defined for the operators used so that the MDD representation remains unique.

    As an example, the work in [69] considers cyclic negation as edge operators. The MDD in this case have a single terminal vertex with value 0 and no cycle operator is permitted on a 0-edge from a vertex. These rules are sufficient to ensure the uniqueness of the representation. The rules are different from the CUDD normalization rules described above because the MDD package is applicable for varying values of p and it is easier to associate the rules with 0 which is the value consistent to all problems.

3.5.3.4 Unique table. In decision diagram packages, it is common to use a unique table to avoid creating multiple instances of a vertex. One straightforward approach is to use a separate subtable for each variable with a fixed number of slots in each subtable. When a vertex is needed, a hash function is applied to associate the vertex with a particular slot in the subtable associated with the vertex selection variable. Each slot in the hash table heads a list (initially empty) of vertices. When a vertex is required, the list for the slot the vertex hashes to is linearly searched. If the vertex is found, the instance found is used. Otherwise, the vertex is added to the beginning of the slot list for possible further use. Experiment has shown that adding to the front of the list is most effective due to the local computation nature of decision diagram construction. Use of a unique table ensures that only one instance of a node is ever created which deals with the need to identify isomorphic subgraphs.

Performance of the unique table in terms of distributing vertices across the slots and thereby reducing the amount of list searching depends on the number of slots and the hash function used. An effective hash function simply sums the shifted pointer addresses (treated as integers) associated with the vertex's outgoing edges and then takes that sum mod the number of slots as the slot index. This is expressed as

$$index = (\Sigma_{i=0}^{p^2-1} a_i \gg i) \ mod_{nslots}$$

where the a_i are the pointer addresses and $x \gg i$ denotes shifting x by i bits to the right with the shifted bits being dropped. The reason for the shifting is to make the order of the edges significant since many vertices have similar edges in different permutations. Experiments have shown that without this shifting, the vertices are not as evenly distributed across the unique table slots resulting in unnecessarily long lists and excessive search times. The *mod* operation, which is computationally expensive, can be replaced by a much more efficient bitwise logical AND if the number of slots (*nslots*) is a power of 2.

3.5.3.5 Computed table. It is common for decision diagram packages to use a computed table [11] to reduce the number of duplicate computations. When an operation is to be performed on two diagrams, the computed table is checked to see if the result of the computation is available there before performing the full computation. Since the decision diagram computations are implemented recursively, e.g., the CASE operator (Algorithm 3.1), this check is performed at each recursive step so that results known for subcomputations can also be reused. The check is actually made after dealing with terminal vertex cases since it is faster to compute the terminal vertex cases directly than to check the computed table and not putting terminal cases in the computed table leads to fewer table lookup conflicts.

A typical computed table is a list of slots each of which can hold the operation type and the edges pointing to the operands and the result diagram. When an operation is to be performed, the edges pointing to the operand diagrams are used to choose a slot in the computed table by means of a simple hash function such as adding the two pointer addresses modulo the number of slots in the table. If the identified slot, contains the two operand edges and the desired operation type, the result edge is retrieved from that slot. However, if the slot is empty or contains different operands or a different operation, the required operation is performed and an entry is then made into the chosen slot in the computed table. The entry overwrites a previous entry in the slot.

As for the unique table, note that the expensive mod operation is avoided by having a power of two slots in the table. Also note that each slot is a single entry and not a list as in the unique table. The difference is that it is acceptable to have to repeat a computation whereas it is not acceptable to have two equivalent vertices in the unique table. Hence the rather expensive task of searching lists is required for the unique table, whereas it has been found to not provide any advantage for a computed table. In fact, using lists in the computed table can slow operations down since operations would remain in the table effectively forever as there is no way to tell when they will not be needed again. This leads to very long lists.

3.5.3.6 Reference counts and garbage collection. During the course of constructing and manipulating decision diagrams, including MDD, many vertices become unnecessary since they were either temporary in the midst of a construction, or part of a diagram that is no longer needed. For large problems, it is often necessary to recover the memory for such vertices to avoid unnecessarily excessive memory requirements. The key of course is a simple mechanism to determine if a vertex is no longer in use.

A commonly used approach is to associate a reference count with each vertex which is the number of edges that point to that vertex. The reference count is set to 0 when a vertex is first constructed and is increased/decreased as edges pointing to the vertex are added/removed.

A vertex is no longer in use if its reference count becomes 0. The memory space can be immediately recovered, but that is a fairly costly approach since it can lead to the overhead of recovering memory when it is not needed. Also, it frequently happens that a vertex that passes out of use is required again later. If it has been recycled, it must be reconstructed.

The usual approach is to perform a *garbage collection* operation either initiated by the user, or more commonly, as a dynamically invoked operation when the total memory use exceeds a set limit. A garbage collector scans through all the vertices in the unique table, and collects those with 0 reference count. Collected vertices are removed from the unique table and the associate memory is made available for reuse. This can be done by returning the freed memory to the operating system for later reallocation, but that can involve a fair amount of overhead. An alternative is for the decision diagram package to keep its own list of free space and to get memory via the operating system only when that list is empty. The implementation of garbage collection in CUDD [106] is a very good example for those interested in full detail on this issue.

Garbage collection introduces an issue regarding the computed table since it is possible that the table includes a result edge pointing to a vertex with 0 reference count. When the vertex is collected for reuse, the reference can not remain in the computed table. One solution, is to scan the computed table following garbage collection and to empty all slots that have a result edge pointing to a vertex with 0 reference count.

3.5.3.7 Variable reordering. As for other decision diagram structures, the size (number of vertices) of an MDD is quite dependent on the variable ordering. There are $n!$ orderings for n variables so finding an optimal ordering is not feasible. Sifting [92] only examines on the order of n^2 variable orderings yet yields quite good results. Originally developed for BDD, sifting is readily extended to other forms of decision diagrams and is the most commonly used heuristic. We first present the basic sifting algorithm and then discuss how it can be applied to MDD.

Algorithm 3.2 Sifting Procedure.

1. Select a variable y – a simple heuristic is to choose the variable that labels the most nodes in the decision diagram, choosing the lowest variable in the case of ties.

2. Sift y to the bottom of the decision diagram by a sequence of adjacent variable interchanges.

3. Sift y to the top of the decision diagram by a sequence of adjacent variable interchanges.

4. During steps (2) and (3) a record is kept of the position of y that yields the smallest vertex count in the decision diagram (highest position in the case of ties), so now sift y back down to that position.

5. Repeat steps (1) to (4) until each variable has been sifted into its best position noting that once a variable is selected for sifting, it is not selected a second time.

The key to the above algorithm is adjacent variable interchange which refers to interchanging the order of two variables in the ordering underlying the decision diagram. When such an interchange is made, vertices labeled by those variables must be adjusted accordingly. The key is to do this as a local transformation on those vertices avoiding alteration to other vertices in the diagram.

Algorithm 3.3 MDD Adjacent Variable Interchange.   We consider the case of interchanging variables x and y where the former is immediately above the latter in the MDD.

1. Consider a vertex a with selection variable x. We construct a table T with p rows and p columns, where p is the radix of the function represented by the MDD.

2. For $i = 0, 1, \ldots, p - 1$,

   - If the ith-edge from a leads to a vertex b with selection variable y, then for $j = 0, 1, \ldots, p - 1$, $T_{i,j}$ is set to point to the vertex pointed to by the jth-edge from b with the edge negations being the composition of the edge negations on the ith-edge from a and the jth-edge from b.

   - If the ith-edge from a leads to a vertex b not labeled y, $i.e.$, it points lower in the MDD, then $T_{i,j}$ is set to the ith-edge from a for $j = 0, 1, \ldots, p - 1$.

3. Once T is constructed as above, the level interchange is made by relabeling vertex a with y, and setting each ith-edge from a, $i = 0, 1, \ldots, p - 1$ to point to a vertex labeled x whose jth-edge, $j = 0, 1, \ldots, p - 1$, points to the vertex $T_{j,i}$ points to (note the order of the subscripts which is key to the interchange). During this construction, edge operations are normalized as described above.

4. The complete level interchange is accomplished by performing the above for all vertices originally labeled x.

The idea of relabeling vertices originally labeled x is critical as it means that edges leading to them, and the vertices from which those edges originate, are unaffected by the level interchange. These vertices must be removed from the unique table and reinserted appropriately after the interchange.

It is easily confirmed that after applying this algorithm, every vertex a originally labeled by x and now labeled y, is the top of a MDD representing the same function it did when originally labeled x. The nodes originally labeled y are affected as edges to them are removed. Reference count-based garbage collection must be used as such a vertex can not be discarded unless no other vertices higher in the diagram point to it. Finally, no vertex below the two levels being interchanged is affected except for changing the reference counts. The result is that the adjacent variable interchange is a local operation affecting only the two levels being interchanged and reference counts for some nodes below those levels.

The above technique is a generalization of the method introduced by Rudell [92]. Algorithm 3.3 has been described assuming all variables in the MDD are p-valued. It is readily extended to the case where variables assume different numbers of values [70].

Given the above method for adjacent variable interchange, sifting of MDD is readily implemented using the basic sifting approach described earlier. Note that several extensions and enhancements to basic sifting have been developed. The interested reader should consult the literature.

TABLE 3.3: MDD construction and sifting results [?].

EXAMPLE	P	IN	OUT	INITIAL SIZE	SIFTED SIZE
alu2	2	10	8	114	60
	4	5+	8	90	54
	4	5*	8	48	48
alu4	2	14	8	1093	570
	4	7+	8	786	573
	4	7*	8	408	381
apex1	2	45	45	4876	1307
	4	23+	45	3081	874
	4	23*	45	933	874
apex2	2	39	3	5613	400
	4	20+	3	3470	646
	4	20*	3	299	274
apex3	2	54	50	1044	904
	4	27+	50	597	563
	4	27*	50	766	602
bw	2	5	28	112	99
	4	3+	28	88	69
	4	3*	28	81	63
seq	2	41	35	2153	1201
	4	21+	35	1300	857
	4	21*	35	933	857
e64	2	65	65	1444	129
	4	33+	65	1019	592
	4	33*	65	162	162
duke2	2	22	29	769	369
	4	11+	29	561	339
	4	11*	29	279	278
misex1	2	8	7	71	36
	4	4+	7	47	25
	4	4*	7	33	33

TABLE 3.3 (*Continued*):

EXAMPLE	P	IN	OUT	INITIAL SIZE	SIFTED SIZE
misex2	2	25	18	114	82
	4	13+	18	114	81
	4	13*	18	78	64
misex3	2	14	14	652	480
	4	14+	14	433	321
	4	14*	14	318	317
sao2	2	10	4	126	86
	4	5+	4	81	64
	4	5*	4	56	56

(+ original order conversion; * sifted order conversion)

Experimental results on the construction of MDD using the techniques described in this Chapter can be found in [69,70]. The work in [70] also introduces a technique termed *augmented sifting* which is a generalization of *linear sifting* [68] for BDD.

Example 3.12.   Table 3.3 presents results from [70]. The binary ($p = 2$) functions are from the LGSYNTH91 benchmark set [123]. Each binary function is also converted to four-valued input, binary output problems by adjacent variable pairing. This is done in two ways for each function: pairing based on the variable order in the specification as given, and pairing based on the sifted variable ordering. Not surprisingly the conversion based on the sifted ordering consistently yields smaller MDD. Sifting the MDD yields further improvement in most cases. The MDD as expected have lower vertex counts than the BDD.

CHAPTER 4

Reversible and Quantum Circuits

A computation is reversible if, and only if, there is no information loss. In terms of circuits, this means every possible assignment to the inputs must result in a unique output pattern. Landauer's principle [52] sets a lower bound on the energy dissipation associated with the erasure of a bit of information. Reversible circuits have thus been of interest for some time due to the potential for minimizing energy consumption. Reversibility has, for example, been considered in the design of low-power CMOS circuits [5]. Interest in reversibility has also been heightened by the considerable interest in quantum computation and quantum information, and specifically because quantum gates and circuits are reversible.

The functional behavior of a reversible (including quantum) gate can be represented as a transformation matrix and the functional behavior of a cascade (circuit) composed of such gates is given by the product of those matrices. These matrices are, in general, complex-valued and can be quite large since the dimension is $p^n \times p^n$ for a p-valued n-line circuit. The challenges we aim to address are how to represent and efficiently manipulate these matrices. Our approach is similar to MDD and uses a variety of other decision diagram (DD) techniques.

The presentation of reversible and quantum gates and circuits in this Chapter emphasizes the details required for the development of quantum multiple-valued decision diagram (QMDD) (see Chapter 5). Readers wanting a more in-depth discussion of reversible and quantum circuits, or who are interested in the circuit synthesis problem, are advised to consult one of the many books and articles on those areas, e.g., [1,28,41,48,49,54,61–63,71,73,82,102].

4.1 BINARY REVERSIBLE GATES AND CIRCUITS

While quantum gates and circuits are reversible, we shall follow the convention of using *reversible* to refer to gates and circuits which operate on the values $0, \ldots, p-1$ with $p = 2$ for binary and $p > 2$ for multiple-valued throughout the circuit. This is to clearly distinguish those cases from the quantum case discussed in Section 4.4.

Definition 4.1. An $n \times n$ **binary reversible function** has n inputs and n outputs and is a mapping from $\{0, 1\}^n$ onto $\{0, 1\}^n$.

Since the domain and range are both $\{0, 1\}^n$ and the mapping is *onto*, an $n \times n$ binary reversible function is a permutation of $\{0, 1\}^n$ and can thus be expressed as a $2^n \times 2^n$ permutation matrix P.

The fact the function is reversible means there is a unique inverse function that is defined by the inverse permutation given by P^{-1} which for all permutation matrices is P^t. Construction of the appropriate matrix is discussed below.

The NOT gate is reversible and is also self-inverse. The logic gates AND, OR, NAND, NOR, XOR, and XNOR are not reversible. In fact, it is important to recognize that no symmetric function, including the gates just noted, is reversible. Conversely no reversible function can be symmetric. This is because symmetry, in either all or a subset of the variables, means that some set of distinct input assignments are mapped to a common output pattern, which violates the unique mapping required for reversibility.

A commonly used family of reversible binary gates are the $k \times k$ Toffoli gates, which are a generalization of the NOT gate. The notation $k \times k$ indicates the gate has k inputs and k outputs. Note that the reversible gates and circuits we deal with here are fully defined and consequently always have the same number of inputs and outputs. This is not necessarily the case if partially defined functions are considered.

Definition 4.2. A $k \times k$ **Toffoli gate** passes $k - 1$ lines (controls) through unchanged and inverts the remaining line (target) if the control lines are all 1, otherwise the value on the target lines passes through the gate unaltered.

A 1×1 Toffoli gate has no control lines and is equivalent to a NOT gate. The 2×2 Toffoli gate and the 3×3 Toffoli gate are often referred to as the *controlled-NOT* and *controlled-controlled-NOT* gates, respectively. The 3×3 gate is also referred to as simply a Toffoli gate.

Another binary reversible gate family is the family of $k \times k$ Fredkin gates.

Definition 4.3. An $k \times k$ **Fredkin gate** passes $k - 2$ lines (controls) through unchanged and interchanges the values on the remaining two lines (targets) if the control lines are all 1, otherwise the values on the target lines pass through the gate unaltered.

A 2×2 Fredkin gate, i.e., one with no controls always swaps the values on the two lines and is thus often referred to as a *swap* gate.

Like the NOT gate, Tofolli and Fredkin gates are self-inverse since two consecutive identical gates result in an identity permutation.

Note that the above definitions allow any number of control lines which differs from the initial definitions put forward by Toffoli [15] and Fredkin [33]. They should thus perhaps be termed *generalized* or *multicontrol* gates but we will use the simpler terminology. Also, while some work has permitted the activation value for each control line to be specified as 0 or 1, we will here consider only the 1 case as it is the most prevalent.

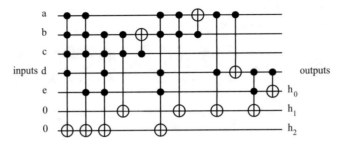

FIGURE 4.1: Reversible circuit for function RD53.

Definition 4.4. An $n \times n$ **binary reversible circuit** is a cascade of binary reversible gates each involving up to n lines.

Being a cascade, a reversible circuit has no fan-out and no feedback loops. It is important to emphasize that each gate involves from 1 to n lines and applies the appropriate operation to the target line(s) if the required control condition is met. Lines unused in a gate are passed through to the next gate in the cascade, as are the control lines.

Example 4.1. Figure 4.1 shows a reversible circuit with 12 Toffoli gates for a commonly used benchmark function RD53 [123]. For each gate, the control connections are marked with black circles. The symbol $\oplus$ is on the target line and denotes the negation performed when the control conditions are met.

RD53 is a 5-input, 3-output function and hence is not itself reversible. It has been made reversible by adding two *constant* inputs set to 0, and by adding four *garbage* outputs which are unlabeled in the diagram. Our convention is to have circuit inputs on the left and circuit outputs on the right. Reading the circuit in the reverse direction is a realization for the inverse function.

Note that in Fig. 4.1, the function RD53 is embedded in a 7×7 reversible function. This is an often required procedure as most functions of practical interest are not themselves reversible. Finding an optimal embedding, i.e., an embedding that leads to a low-cost and efficient circuit, is to date an unsolved problem which is beyond the scope of the discussion here. Heuristic techniques, such as those described in [75], or specific knowledge of the function are used.

4.2 MULTIPLE-VALUED LOGIC REVERSIBLE GATES AND CIRCUITS

Generalizing the binary ideas outlined above to multiple-valued logic (MVL) is reasonably straightforward. Note that there are several possible generalizations of the binary NOT operation and hence several generalizations of Toffoli gates. As in the binary case, each MVL reversible gate and circuit implements a permutation.

TABLE 4.1: Three-valued controlled-cycle-by-1

X_1	X_0	X_1	$CC1(X_1; X_0)$
0	0	0	0
0	1	0	1
0	2	0	2
1	0	1	0
1	1	1	1
1	2	1	2
2	0	2	1
2	1	2	2
2	2	2	0

Consider the MVL unary operators in Table 2.4. *Negation, cycle, successor*, and *predecessor* are reversible, while the *decisive, window, selection*, and *threshold* literals clearly are not. The commonly used two or higher-input MVL gates such as *MAX, MIN, mod-SUM, truncated-SUM*, as well as any other gates implementing full or partially symmetric functions, are not reversible for the reasons discussed earlier.

A. De Vos *et al.* [25] considered ternary cycle (C1) and negation (N) gates and the controlled versions of those gates, denoted CC1 and CN (see Tables 4.1 and 4.2) as generators of the group of all 2×2 three-valued reversible logic functions.

TABLE 4.2: Three-valued controlled negation

X_1	X_0	X_1	$CN(X_1; X_0)$
0	0	0	0
0	1	0	1
0	2	0	2
1	0	1	0
1	1	1	1
1	2	1	2
2	0	2	2
2	1	2	1
2	2	2	0

TABLE 4.3: Minimal gate counts for two-line ternary reversible circuits for various basic gate sets

GATES	C1-CN	C1-CC1-CN	C1-N-CC1-CN	C1-C2-N CC1-CC2-CN
0	1	1	1	1
1	4	6	8	12
2	13	31	52	93
3	39	130	280	597
4	115	498	1,342	3,224
5	326	1,777	5,692	15,042
6	897	5,924	20,992	57,951
7	2,395	18,089	63,292	144,039
8	6,107	47,849	128,159	127,056
9	14,660	99,576	118,635	14,750
10	32,268	126,981	23,516	115
11	62,145	58,192	906	
12	96,237	3,795	5	
13	97,705	31		
14	43,902			
15	5,816			
16	243			
17	7			
Avg. Gates	11.97	9.39	8.11	7.16

A computer program [71] designed to examine the size (number of gates) of circuits realizing the $9! = 326, 880$ 2×2 three-valued reversible logic functions for different sets of basic operations gave the results in Table 4.3. The program performs a breadth-first search of all possible circuits until a realization for every function has been found. Since the search is breadth-first, the circuit found for each function uses a minimal number of gates.

Cycle (C1) together with controlled negation (CN) can realize all 326,880 functions. The circuits can be quite long. Adding, controlled cycle (CC1) and negation (N) in turn both improve the results. The latter is the set of operators suggested by De Vos *et al.* [25].

The rightmost column in Table 4.3 shows the further improvement gained by adding cycle by 2 (C2) and controlled cycle by 2 (CC2). C1, C2, and N are extensions of the notion of binary inversion and are clearly reversible. N is self-inverse and C1 and C2 are the inverse of each other. CC1, CC2, and CN are extensions of the 2×2 Toffoli gate and are each readily seen to be reversible.

Negation extends to any p-valued logic with $x' = (p - 1) - x$. There are $p - 1$ cycle inversions for p-valued logic defined in the obvious way. Controlled cycle and controlled negation generalize to the $k \times k$ cases for p-valued logic in the following two definitions.

Definition 4.5.   A $k \times k$ p-valued **controlled cycle gate** passes $k - 1$ lines (controls) through unchanged and cycles the remaining line (target) by a specified amount (1 to $p - 1$) if every control line has its appropriate activation value, otherwise the value on the target line passes through the gate unaltered.

Definition 4.6.   A $k \times k$ p-valued **controlled negation gate** passes $k - 1$ lines (controls) through unchanged and negates the remaining line (target) if every control line has its appropriate activation value, otherwise the value on the target line passes through the gate unaltered.

Note that in the above definitions, the activation values for the control lines do not have to be the same but the activation value for each control line in a gate is a single fixed value. A further generalization, not considered here, is to allow a control to be activated by a proper nonempty subset of the values $0, \ldots, p - 1$.

Fredkin gates can be generalized to the MVL case in the obvious and straightforward way. We leave that to the consideration of the reader.

A MVL reversible circuit is a cascade of reversible MVL gates with no fan-out or feedback loops.

Example 4.2.   Figure 4.2 is a reversible implementation of a ternary full adder produced by the synthesis procedure reported in [71]. The white squares represent target line operations labeled by the appropriate function. The black circles represent controls and are labeled by the appropriate control activation values.

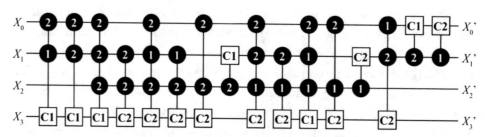

FIGURE 4.2: Reversible circuit for ternary full adder.

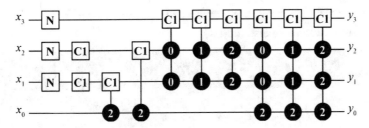

FIGURE 4.3: Reversible circuit for multiplicative inverse with $p = 3$ and $k = 4$.

The next example is based on the following definition:

Definition 4.7.   The multiplicative inverse modulo-p^k of a fixed-point value a, denoted a^{-1}, satisfies

$$1 = (a \times a^{-1})(mod\ p^k)$$

where both a and a^{-1} are expressed as digit strings of length k with $(a, a^{-1}) \in Z$.

Example 4.3.   For $p = 3$ and $k = 4$, we have the function specified in Table 4.4. It can be shown that this function is realized by the circuit in Fig. 4.3.

4.3   MATRIX REPRESENTATION OF REVERSIBLE GATES AND CIRCUITS

The transformation performed by a binary, $p = 2$, or multiple-valued, $p > 2$, $k \times k$ reversible gate can be represented as a permutation matrix of dimension $p^k \times p^k$. We present illustrative examples followed by a rigorous method for constructing the matrices.

Example 4.4.   For $k = 3, p = 2$, the Toffoli gate $T(x_0, x_2; x_1)$ with controls x_0, x_2 and target x_1 (see Fig. 4.4) has the truth table in Table 4.5.

The corresponding permutation matrix is shown in Eq. 4.1.

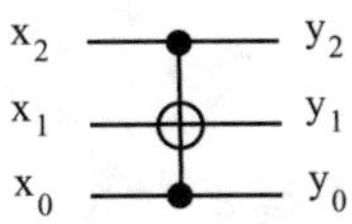

$$
\begin{array}{c}
x_2 \quad\longrightarrow\quad y_2 \\
x_1 \quad\longrightarrow\quad y_1 \\
x_0 \quad\longrightarrow\quad y_0
\end{array}
$$

FIGURE 4.4: $T(x_0, x_2; x_1)$ gate.

TABLE 4.4: Multiplicative inverse mod 3^4

A (DECIMAL)	A^{-1} (DECIMAL)	A (TERNARY)	A^{-1} (TERNARY)
1	1	0001	0001
2	41	0002	1112
4	61	0011	2021
5	65	0012	2102
7	58	0021	2011
8	71	0022	2122
10	73	0101	2201
11	59	0102	2012
13	25	0111	0221
14	29	0112	1002
16	76	0121	2211
17	62	0122	2022
19	64	0201	2101
20	77	0202	2212
22	70	0211	2121
23	74	0212	2202
25	13	0221	0111
26	53	0222	1222
28	55	1001	2001
29	25	1002	0112
31	34	1011	1021
32	38	1012	1102
34	31	1021	1011
35	44	1022	1122
37	46	1101	1201
38	32	1102	1012
40	79	1111	2221
41	2	1112	0002
43	49	1121	1211
44	35	1122	1022
46	37	1201	1101
47	50	1202	1212

TABLE 4.4: (*Continued*)

A (DECIMAL)	A⁻¹ (DECIMAL)	A (TERNARY)	A⁻¹ (TERNARY)
49	43	1211	1121
50	47	1212	1202
52	67	1221	2111
53	26	1222	0222
55	28	2001	1001
56	68	2002	2112
58	7	2011	0021
59	11	2012	0102
61	4	2021	0011
62	17	2022	0122
64	19	2101	0201
65	5	2102	0012
67	52	2111	1221
68	56	2112	2002
70	22	2121	0211
71	8	2122	0022
73	10	2201	0101
74	23	2202	0212
76	16	2211	0121
77	20	2212	0202
79	40	2221	1111
80	80	2222	2222

$$P = \begin{bmatrix} 1 & 0 & 0 & 0 & 0 & 0 & 0 & 0 \\ 0 & 1 & 0 & 0 & 0 & 0 & 0 & 0 \\ 0 & 0 & 1 & 0 & 0 & 0 & 0 & 0 \\ 0 & 0 & 0 & 1 & 0 & 0 & 0 & 0 \\ 0 & 0 & 0 & 0 & 1 & 0 & 0 & 0 \\ 0 & 0 & 0 & 0 & 0 & 0 & 0 & 1 \\ 0 & 0 & 0 & 0 & 0 & 0 & 1 & 0 \\ 0 & 0 & 0 & 0 & 0 & 1 & 0 & 0 \end{bmatrix} \qquad (4.1)$$

TABLE 4.5: Truth table for $T(x_0, x_2; x_1)$

X_2	X_1	X_0	Y_2	Y_1	Y_0
0	0	0	0	0	0
0	0	1	0	0	1
0	1	0	0	1	0
0	1	1	0	1	1
1	0	0	1	0	0
1	0	1	1	1	1
1	1	0	1	1	0
1	1	1	1	0	1

Equation 4.2 verifies the correctness of the permutation matrix shown in Eq. 4.1 and illustrates its application.

$$
\begin{array}{|ccc|}
x_2 & x_1 & x_0 \\
\hline
0 & 0 & 0 \\
0 & 0 & 1 \\
0 & 1 & 0 \\
0 & 1 & 1 \\
1 & 0 & 0 \\
1 & 1 & 1 \\
1 & 1 & 0 \\
1 & 0 & 1
\end{array}
=
\begin{bmatrix}
1 & 0 & 0 & 0 & 0 & 0 & 0 & 0 \\
0 & 1 & 0 & 0 & 0 & 0 & 0 & 0 \\
0 & 0 & 1 & 0 & 0 & 0 & 0 & 0 \\
0 & 0 & 0 & 1 & 0 & 0 & 0 & 0 \\
0 & 0 & 0 & 0 & 1 & 0 & 0 & 0 \\
0 & 0 & 0 & 0 & 0 & 0 & 0 & 1 \\
0 & 0 & 0 & 0 & 0 & 0 & 1 & 0 \\
0 & 0 & 0 & 0 & 0 & 1 & 0 & 0
\end{bmatrix}
\times
\begin{array}{|ccc|}
x_2 & x_1 & x_0 \\
\hline
0 & 0 & 0 \\
0 & 0 & 1 \\
0 & 1 & 0 \\
0 & 1 & 1 \\
1 & 0 & 0 \\
1 & 0 & 1 \\
1 & 1 & 0 \\
1 & 1 & 1
\end{array}
\qquad (4.2)
$$

For a p-valued $k \times k$ reversible gate, the transformation matrix is a $p^k \times p^k$ permutation matrix.

Example 4.5. Consider a four-valued controlled-negation (CN) gate $CN(x_1[2]; x_0)$ with control x_1 with activation value 2, and target x_0 has the truth table in Table 4.6. The permutation matrix is shown in Eq. 4.3.

$$P = \begin{bmatrix}
1 & 0 & 0 & 0 & 0 & 0 & 0 & 0 & 0 & 0 & 0 & 0 & 0 & 0 & 0 & 0 \\
0 & 1 & 0 & 0 & 0 & 0 & 0 & 0 & 0 & 0 & 0 & 0 & 0 & 0 & 0 & 0 \\
0 & 0 & 1 & 0 & 0 & 0 & 0 & 0 & 0 & 0 & 0 & 0 & 0 & 0 & 0 & 0 \\
0 & 0 & 0 & 1 & 0 & 0 & 0 & 0 & 0 & 0 & 0 & 0 & 0 & 0 & 0 & 0 \\
0 & 0 & 0 & 0 & 1 & 0 & 0 & 0 & 0 & 0 & 0 & 0 & 0 & 0 & 0 & 0 \\
0 & 0 & 0 & 0 & 0 & 1 & 0 & 0 & 0 & 0 & 0 & 0 & 0 & 0 & 0 & 0 \\
0 & 0 & 0 & 0 & 0 & 0 & 1 & 0 & 0 & 0 & 0 & 0 & 0 & 0 & 0 & 0 \\
0 & 0 & 0 & 0 & 0 & 0 & 0 & 1 & 0 & 0 & 0 & 0 & 0 & 0 & 0 & 0 \\
0 & 0 & 0 & 0 & 0 & 0 & 0 & 0 & 0 & 0 & 0 & 1 & 0 & 0 & 0 & 0 \\
0 & 0 & 0 & 0 & 0 & 0 & 0 & 0 & 0 & 0 & 1 & 0 & 0 & 0 & 0 & 0 \\
0 & 0 & 0 & 0 & 0 & 0 & 0 & 0 & 0 & 1 & 0 & 0 & 0 & 0 & 0 & 0 \\
0 & 0 & 0 & 0 & 0 & 0 & 0 & 0 & 1 & 0 & 0 & 0 & 0 & 0 & 0 & 0 \\
0 & 0 & 0 & 0 & 0 & 0 & 0 & 0 & 0 & 0 & 0 & 0 & 1 & 0 & 0 & 0 \\
0 & 0 & 0 & 0 & 0 & 0 & 0 & 0 & 0 & 0 & 0 & 0 & 0 & 1 & 0 & 0 \\
0 & 0 & 0 & 0 & 0 & 0 & 0 & 0 & 0 & 0 & 0 & 0 & 0 & 0 & 1 & 0 \\
0 & 0 & 0 & 0 & 0 & 0 & 0 & 0 & 0 & 0 & 0 & 0 & 0 & 0 & 0 & 1
\end{bmatrix} \tag{4.3}$$

TABLE 4.6: Truth table for four-valued $CN(x_1[2]; x_0)$

X_1	X_0	Y_1	Y_0
0	0	0	0
0	1	0	1
0	2	0	2
0	3	0	3
1	0	1	0
1	1	1	1
1	2	1	2
1	3	1	3
2	0	2	3
2	1	2	2
2	2	2	1
2	3	2	0

TABLE 4.6: (*Continued*)

X_1	X_0	Y_1	Y_0
3	0	3	0
3	1	3	1
3	2	3	2
3	3	3	3

Each type of reversible gate has a matrix defining the operation (permutation) to be performed on the target line(s). Some examples are shown in Table 4.7. Algorithm 4.1 gives a procedure for constructing the permutation matrix for a p-valued reversible gate with j target lines in an n-line circuit.

Algorithm. Reversible Gate Matrix Construction

1. Assume the target lines are $x_0, \ldots, x_{j-1}$, and begin by setting M_{j-1} to the appropriate target matrix.

2. For i from j to $n-1$ consider each x_i and construct M_i which has dimension $p^{i+1} \times p^{i+1}$ as follows:

   (a) If x_i is an unconnected line for the gate, it has no influence on the permutation to be applied by the gate. Hence

$$M_i = \begin{bmatrix} M_{i-1} & & & \\ & M_{i-1} & & \\ & & \ddots & \\ & & & M_{i-1} \end{bmatrix} \qquad (4.4)$$

where there are p blocks on the diagonal and all entries outside the blocks on the diagonal are 0.

TABLE 4.7: Target operation matrices for selected reversible gates.

TYPE	P	MATRIX
Toffoli	2	$\begin{bmatrix} 0 & 1 \\ 1 & 0 \end{bmatrix}$
Fredkin	2	$\begin{bmatrix} 1 & 0 & 0 & 0 \\ 0 & 0 & 1 & 0 \\ 0 & 1 & 0 & 0 \\ 0 & 0 & 0 & 1 \end{bmatrix}$
N	3	$\begin{bmatrix} 0 & 0 & 1 \\ 0 & 1 & 0 \\ 1 & 0 & 0 \end{bmatrix}$
C1	3	$\begin{bmatrix} 0 & 1 & 0 \\ 0 & 0 & 1 \\ 1 & 0 & 0 \end{bmatrix}$
C2	3	$\begin{bmatrix} 0 & 0 & 1 \\ 1 & 0 & 0 \\ 0 & 1 & 0 \end{bmatrix}$
N	4	$\begin{bmatrix} 0 & 0 & 0 & 1 \\ 0 & 0 & 1 & 0 \\ 0 & 1 & 0 & 0 \\ 1 & 0 & 0 & 0 \end{bmatrix}$
C1	4	$\begin{bmatrix} 0 & 1 & 0 & 0 \\ 0 & 0 & 1 & 0 \\ 0 & 0 & 0 & 1 \\ 1 & 0 & 0 & 0 \end{bmatrix}$
C2	4	$\begin{bmatrix} 0 & 0 & 1 & 0 \\ 0 & 0 & 0 & 1 \\ 1 & 0 & 0 & 0 \\ 0 & 1 & 0 & 0 \end{bmatrix}$

TABLE 4.7: (*Continued*)

TYPE	P	MATRIX
C3	4	$\begin{bmatrix} 0 & 0 & 0 & 1 \\ 1 & 0 & 0 & 0 \\ 0 & 1 & 0 & 0 \\ 0 & 0 & 1 & 0 \end{bmatrix}$

(b) If x_i is a control line with activation value v, M_i is given by:

$$M_i = \begin{bmatrix} I^{i-1} \\ & \ddots \\ & & I^{i-1} \\ & & & M_{i-1} \\ & & & & I^{i-1} \\ & & & & & \ddots \\ & & & & & & I^{i-1} \end{bmatrix}$$

where I^{i-1} denotes an identity matrix of dimension $p^i \times p^i$ and again all entries outside the p blocks on the diagonal are 0. The M_{i-1} block on the diagonal is positioned in the vth position counting from 0 from the top left block.

3. M_{n-1}, which has dimension $p^n \times p^n$, is the permutation describing the operation of the gate with all n lines, targets, controls, and unconnected lines, taken into account.

Example 4.6. Consider building the matrix for the Toffoli gate $T(x_1, x_3; x_0)$ in a four-line circuit. The matrix is constructed as follows (note that in the block matrices, the 0 blocks off the diagonal are omitted):

1. Since the target line is x_0, $M_0 = \begin{bmatrix} 0 & 1 \\ 1 & 0 \end{bmatrix}$.

2. x_1 is a control line so $M - 1 = \begin{bmatrix} I^0 \\ & M_0 \end{bmatrix} = \begin{bmatrix} 1 & 0 & 0 & 0 \\ 0 & 1 & 0 & 0 \\ 0 & 0 & 0 & 1 \\ 0 & 0 & 1 & 0 \end{bmatrix}$

3. x_2 is not connected so

$$M_2 = \begin{bmatrix} M_1 & \\ & M_1 \end{bmatrix} = \begin{bmatrix} 1 & 0 & 0 & 0 & 0 & 0 & 0 & 0 \\ 0 & 1 & 0 & 0 & 0 & 0 & 0 & 0 \\ 0 & 0 & 0 & 1 & 0 & 0 & 0 & 0 \\ 0 & 0 & 1 & 0 & 0 & 0 & 0 & 0 \\ 0 & 0 & 0 & 0 & 1 & 0 & 0 & 0 \\ 0 & 0 & 0 & 0 & 0 & 1 & 0 & 0 \\ 0 & 0 & 0 & 0 & 0 & 0 & 0 & 1 \\ 0 & 0 & 0 & 0 & 0 & 0 & 1 & 0 \end{bmatrix}$$

4. Lastly, since x_3 is a control,

$$M_3 = \begin{bmatrix} I^2 & \\ & M_2 \end{bmatrix} = \begin{bmatrix} 1 & 0 & 0 & 0 & 0 & 0 & 0 & 0 & 0 & 0 & 0 & 0 & 0 & 0 & 0 & 0 \\ 0 & 1 & 0 & 0 & 0 & 0 & 0 & 0 & 0 & 0 & 0 & 0 & 0 & 0 & 0 & 0 \\ 0 & 0 & 1 & 0 & 0 & 0 & 0 & 0 & 0 & 0 & 0 & 0 & 0 & 0 & 0 & 0 \\ 0 & 0 & 0 & 1 & 0 & 0 & 0 & 0 & 0 & 0 & 0 & 0 & 0 & 0 & 0 & 0 \\ 0 & 0 & 0 & 0 & 1 & 0 & 0 & 0 & 0 & 0 & 0 & 0 & 0 & 0 & 0 & 0 \\ 0 & 0 & 0 & 0 & 0 & 1 & 0 & 0 & 0 & 0 & 0 & 0 & 0 & 0 & 0 & 0 \\ 0 & 0 & 0 & 0 & 0 & 0 & 1 & 0 & 0 & 0 & 0 & 0 & 0 & 0 & 0 & 0 \\ 0 & 0 & 0 & 0 & 0 & 0 & 0 & 1 & 0 & 0 & 0 & 0 & 0 & 0 & 0 & 0 \\ 0 & 0 & 0 & 0 & 0 & 0 & 0 & 0 & 1 & 0 & 0 & 0 & 0 & 0 & 0 & 0 \\ 0 & 0 & 0 & 0 & 0 & 0 & 0 & 0 & 0 & 1 & 0 & 0 & 0 & 0 & 0 & 0 \\ 0 & 0 & 0 & 0 & 0 & 0 & 0 & 0 & 0 & 0 & 0 & 1 & 0 & 0 & 0 & 0 \\ 0 & 0 & 0 & 0 & 0 & 0 & 0 & 0 & 0 & 0 & 1 & 0 & 0 & 0 & 0 & 0 \\ 0 & 0 & 0 & 0 & 0 & 0 & 0 & 0 & 0 & 0 & 0 & 0 & 1 & 0 & 0 & 0 \\ 0 & 0 & 0 & 0 & 0 & 0 & 0 & 0 & 0 & 0 & 0 & 0 & 0 & 1 & 0 & 0 \\ 0 & 0 & 0 & 0 & 0 & 0 & 0 & 0 & 0 & 0 & 0 & 0 & 0 & 0 & 0 & 1 \\ 0 & 0 & 0 & 0 & 0 & 0 & 0 & 0 & 0 & 0 & 0 & 0 & 0 & 0 & 1 & 0 \end{bmatrix}$$

In the above procedure, we assume the target lines are $x_0, \ldots, x_{j-1}$. We can clearly move the target line(s) to occur elsewhere by permuting the rows and columns of M_{n-1} appropriately.

For example, suppose we want to build the matrix for $T(x_0, x_3; x_1)$ again in a four-line circuit, we can start by building the matrix M for $T(x_1, x_3; x_0)$ as above. The matrix for the desired gate is then given by $\widehat{M} = P \times M \times P^{-1}$ where P is a permutation matrix that interchanges the rows and columns of a truth table consistent with interchanging x_0 and x_1. In this case,

$$P = \begin{bmatrix}
1 & 0 & 0 & 0 & 0 & 0 & 0 & 0 & 0 & 0 & 0 & 0 & 0 & 0 & 0 & 0 \\
0 & 0 & 1 & 0 & 0 & 0 & 0 & 0 & 0 & 0 & 0 & 0 & 0 & 0 & 0 & 0 \\
0 & 1 & 0 & 0 & 0 & 0 & 0 & 0 & 0 & 0 & 0 & 0 & 0 & 0 & 0 & 0 \\
0 & 0 & 0 & 1 & 0 & 0 & 0 & 0 & 0 & 0 & 0 & 0 & 0 & 0 & 0 & 0 \\
0 & 0 & 0 & 0 & 1 & 0 & 0 & 0 & 0 & 0 & 0 & 0 & 0 & 0 & 0 & 0 \\
0 & 0 & 0 & 0 & 0 & 0 & 1 & 0 & 0 & 0 & 0 & 0 & 0 & 0 & 0 & 0 \\
0 & 0 & 0 & 0 & 0 & 1 & 0 & 0 & 0 & 0 & 0 & 0 & 0 & 0 & 0 & 0 \\
0 & 0 & 0 & 0 & 0 & 0 & 0 & 1 & 0 & 0 & 0 & 0 & 0 & 0 & 0 & 0 \\
0 & 0 & 0 & 0 & 0 & 0 & 0 & 0 & 1 & 0 & 0 & 0 & 0 & 0 & 0 & 0 \\
0 & 0 & 0 & 0 & 0 & 0 & 0 & 0 & 0 & 0 & 1 & 0 & 0 & 0 & 0 & 0 \\
0 & 0 & 0 & 0 & 0 & 0 & 0 & 0 & 0 & 1 & 0 & 0 & 0 & 0 & 0 & 0 \\
0 & 0 & 0 & 0 & 0 & 0 & 0 & 0 & 0 & 0 & 0 & 1 & 0 & 0 & 0 & 0 \\
0 & 0 & 0 & 0 & 0 & 0 & 0 & 0 & 0 & 0 & 0 & 0 & 1 & 0 & 0 & 0 \\
0 & 0 & 0 & 0 & 0 & 0 & 0 & 0 & 0 & 0 & 0 & 0 & 0 & 0 & 1 & 0 \\
0 & 0 & 0 & 0 & 0 & 0 & 0 & 0 & 0 & 0 & 0 & 0 & 0 & 1 & 0 & 0 \\
0 & 0 & 0 & 0 & 0 & 0 & 0 & 0 & 0 & 0 & 0 & 0 & 0 & 0 & 0 & 1
\end{bmatrix}$$

P is symmetric and $P^{-1} = P$. $\widehat{M}$, the matrix for $T(x_0, x_3; x_1)$ is thus

$$\widehat{M} = P \times M \times P = \begin{bmatrix} 1 & 0 & 0 & 0 & 0 & 0 & 0 & 0 & 0 & 0 & 0 & 0 & 0 & 0 & 0 & 0 \\ 0 & 1 & 0 & 0 & 0 & 0 & 0 & 0 & 0 & 0 & 0 & 0 & 0 & 0 & 0 & 0 \\ 0 & 0 & 1 & 0 & 0 & 0 & 0 & 0 & 0 & 0 & 0 & 0 & 0 & 0 & 0 & 0 \\ 0 & 0 & 0 & 1 & 0 & 0 & 0 & 0 & 0 & 0 & 0 & 0 & 0 & 0 & 0 & 0 \\ 0 & 0 & 0 & 0 & 1 & 0 & 0 & 0 & 0 & 0 & 0 & 0 & 0 & 0 & 0 & 0 \\ 0 & 0 & 0 & 0 & 0 & 1 & 0 & 0 & 0 & 0 & 0 & 0 & 0 & 0 & 0 & 0 \\ 0 & 0 & 0 & 0 & 0 & 0 & 1 & 0 & 0 & 0 & 0 & 0 & 0 & 0 & 0 & 0 \\ 0 & 0 & 0 & 0 & 0 & 0 & 0 & 1 & 0 & 0 & 0 & 0 & 0 & 0 & 0 & 0 \\ 0 & 0 & 0 & 0 & 0 & 0 & 0 & 0 & 1 & 0 & 0 & 0 & 0 & 0 & 0 & 0 \\ 0 & 0 & 0 & 0 & 0 & 0 & 0 & 0 & 0 & 0 & 0 & 1 & 0 & 0 & 0 & 0 \\ 0 & 0 & 0 & 0 & 0 & 0 & 0 & 0 & 0 & 0 & 1 & 0 & 0 & 0 & 0 & 0 \\ 0 & 0 & 0 & 0 & 0 & 0 & 0 & 0 & 0 & 1 & 0 & 0 & 0 & 0 & 0 & 0 \\ 0 & 0 & 0 & 0 & 0 & 0 & 0 & 0 & 0 & 0 & 0 & 0 & 1 & 0 & 0 & 0 \\ 0 & 0 & 0 & 0 & 0 & 0 & 0 & 0 & 0 & 0 & 0 & 0 & 0 & 0 & 0 & 1 \\ 0 & 0 & 0 & 0 & 0 & 0 & 0 & 0 & 0 & 0 & 0 & 0 & 0 & 0 & 1 & 0 \\ 0 & 0 & 0 & 0 & 0 & 0 & 0 & 0 & 0 & 0 & 0 & 0 & 0 & 1 & 0 & 0 \end{bmatrix}$$

Given the above approach to building the matrices for individual gates, we now turn our attention to circuits. Consider an n-line, p-valued circuit consisting of the cascade of reversible gates (from input to output) $G_0, G_1, \ldots, G_{t-1}$, and let M_i denote the $p^n \times p^n$ permutation matrix for G_i. The permutation matrix for the circuit is the product of these matrices as shown in Eq. 4.5. Note the order of the matrix multiplications applies the appropriate transformations from input to output.

$$M = M_{t-1} \times \cdots \times M_1 \times M_0 \qquad (4.5)$$

Example 4.7. Consider the circuit in Fig. 4.5 which is a reversible implementation of a binary full adder [75]. The gates, from left to right, have the permutation matrices:

$$M_0 = \begin{bmatrix}
1 & 0 & 0 & 0 & 0 & 0 & 0 & 0 & 0 & 0 & 0 & 0 & 0 & 0 & 0 & 0 \\
0 & 1 & 0 & 0 & 0 & 0 & 0 & 0 & 0 & 0 & 0 & 0 & 0 & 0 & 0 & 0 \\
0 & 0 & 1 & 0 & 0 & 0 & 0 & 0 & 0 & 0 & 0 & 0 & 0 & 0 & 0 & 0 \\
0 & 0 & 0 & 0 & 0 & 0 & 0 & 0 & 0 & 0 & 0 & 1 & 0 & 0 & 0 & 0 \\
0 & 0 & 0 & 0 & 1 & 0 & 0 & 0 & 0 & 0 & 0 & 0 & 0 & 0 & 0 & 0 \\
0 & 0 & 0 & 0 & 0 & 1 & 0 & 0 & 0 & 0 & 0 & 0 & 0 & 0 & 0 & 0 \\
0 & 0 & 0 & 0 & 0 & 0 & 1 & 0 & 0 & 0 & 0 & 0 & 0 & 0 & 0 & 0 \\
0 & 0 & 0 & 0 & 0 & 0 & 0 & 0 & 0 & 0 & 0 & 0 & 0 & 0 & 0 & 1 \\
0 & 0 & 0 & 0 & 0 & 0 & 0 & 0 & 1 & 0 & 0 & 0 & 0 & 0 & 0 & 0 \\
0 & 0 & 0 & 0 & 0 & 0 & 0 & 0 & 0 & 1 & 0 & 0 & 0 & 0 & 0 & 0 \\
0 & 0 & 0 & 0 & 0 & 0 & 0 & 0 & 0 & 0 & 1 & 0 & 0 & 0 & 0 & 0 \\
0 & 0 & 0 & 1 & 0 & 0 & 0 & 0 & 0 & 0 & 0 & 0 & 0 & 0 & 0 & 0 \\
0 & 0 & 0 & 0 & 0 & 0 & 0 & 0 & 0 & 0 & 0 & 0 & 1 & 0 & 0 & 0 \\
0 & 0 & 0 & 0 & 0 & 0 & 0 & 0 & 0 & 0 & 0 & 0 & 0 & 1 & 0 & 0 \\
0 & 0 & 0 & 0 & 0 & 0 & 0 & 0 & 0 & 0 & 0 & 0 & 0 & 0 & 1 & 0 \\
0 & 0 & 0 & 0 & 0 & 0 & 0 & 1 & 0 & 0 & 0 & 0 & 0 & 0 & 0 & 0
\end{bmatrix}$$

$$M_1 = \begin{bmatrix}
1 & 0 & 0 & 0 & 0 & 0 & 0 & 0 & 0 & 0 & 0 & 0 & 0 & 0 & 0 & 0 \\
0 & 0 & 0 & 1 & 0 & 0 & 0 & 0 & 0 & 0 & 0 & 0 & 0 & 0 & 0 & 0 \\
0 & 0 & 1 & 0 & 0 & 0 & 0 & 0 & 0 & 0 & 0 & 0 & 0 & 0 & 0 & 0 \\
0 & 1 & 0 & 0 & 0 & 0 & 0 & 0 & 0 & 0 & 0 & 0 & 0 & 0 & 0 & 0 \\
0 & 0 & 0 & 0 & 1 & 0 & 0 & 0 & 0 & 0 & 0 & 0 & 0 & 0 & 0 & 0 \\
0 & 0 & 0 & 0 & 0 & 0 & 0 & 1 & 0 & 0 & 0 & 0 & 0 & 0 & 0 & 0 \\
0 & 0 & 0 & 0 & 0 & 0 & 1 & 0 & 0 & 0 & 0 & 0 & 0 & 0 & 0 & 0 \\
0 & 0 & 0 & 0 & 0 & 1 & 0 & 0 & 0 & 0 & 0 & 0 & 0 & 0 & 0 & 0 \\
0 & 0 & 0 & 0 & 0 & 0 & 0 & 0 & 1 & 0 & 0 & 0 & 0 & 0 & 0 & 0 \\
0 & 0 & 0 & 0 & 0 & 0 & 0 & 0 & 0 & 0 & 0 & 0 & 1 & 0 & 0 & 0 \\
0 & 0 & 0 & 0 & 0 & 0 & 0 & 0 & 0 & 0 & 1 & 0 & 0 & 0 & 0 & 0 \\
0 & 0 & 0 & 0 & 0 & 0 & 0 & 0 & 0 & 1 & 0 & 0 & 0 & 0 & 0 & 0 \\
0 & 0 & 0 & 0 & 0 & 0 & 0 & 0 & 0 & 0 & 0 & 0 & 0 & 1 & 0 & 0 \\
0 & 0 & 0 & 0 & 0 & 0 & 0 & 0 & 0 & 0 & 0 & 0 & 0 & 0 & 0 & 1 \\
0 & 0 & 0 & 0 & 0 & 0 & 0 & 0 & 0 & 0 & 0 & 0 & 0 & 0 & 1 & 0 \\
0 & 0 & 0 & 0 & 0 & 0 & 0 & 0 & 0 & 0 & 0 & 0 & 0 & 1 & 0 & 0
\end{bmatrix}$$

$$M_2 = \begin{bmatrix}
1 & 0 & 0 & 0 & 0 & 0 & 0 & 0 & 0 & 0 & 0 & 0 & 0 & 0 & 0 & 0 \\
0 & 1 & 0 & 0 & 0 & 0 & 0 & 0 & 0 & 0 & 0 & 0 & 0 & 0 & 0 & 0 \\
0 & 0 & 1 & 0 & 0 & 0 & 0 & 0 & 0 & 0 & 0 & 0 & 0 & 0 & 0 & 0 \\
0 & 0 & 0 & 1 & 0 & 0 & 0 & 0 & 0 & 0 & 0 & 0 & 0 & 0 & 0 & 0 \\
0 & 0 & 0 & 0 & 1 & 0 & 0 & 0 & 0 & 0 & 0 & 0 & 0 & 0 & 0 & 0 \\
0 & 0 & 0 & 0 & 0 & 1 & 0 & 0 & 0 & 0 & 0 & 0 & 0 & 0 & 0 & 0 \\
0 & 0 & 0 & 0 & 0 & 0 & 0 & 0 & 0 & 0 & 0 & 0 & 0 & 0 & 1 & 0 \\
0 & 0 & 0 & 0 & 0 & 0 & 0 & 0 & 0 & 0 & 0 & 0 & 0 & 0 & 0 & 1 \\
0 & 0 & 0 & 0 & 0 & 0 & 0 & 0 & 1 & 0 & 0 & 0 & 0 & 0 & 0 & 0 \\
0 & 0 & 0 & 0 & 0 & 0 & 0 & 0 & 0 & 1 & 0 & 0 & 0 & 0 & 0 & 0 \\
0 & 0 & 0 & 0 & 0 & 0 & 0 & 0 & 0 & 0 & 1 & 0 & 0 & 0 & 0 & 0 \\
0 & 0 & 0 & 0 & 0 & 0 & 0 & 0 & 0 & 0 & 0 & 1 & 0 & 0 & 0 & 0 \\
0 & 0 & 0 & 0 & 0 & 0 & 0 & 0 & 0 & 0 & 0 & 0 & 1 & 0 & 0 & 0 \\
0 & 0 & 0 & 0 & 0 & 0 & 0 & 0 & 0 & 0 & 0 & 0 & 0 & 1 & 0 & 0 \\
0 & 0 & 0 & 0 & 0 & 0 & 1 & 0 & 0 & 0 & 0 & 0 & 0 & 0 & 0 & 0 \\
0 & 0 & 0 & 0 & 0 & 0 & 0 & 1 & 0 & 0 & 0 & 0 & 0 & 0 & 0 & 0
\end{bmatrix}$$

$$M_3 = \begin{bmatrix}
1 & 0 & 0 & 0 & 0 & 0 & 0 & 0 & 0 & 0 & 0 & 0 & 0 & 0 & 0 & 0 \\
0 & 1 & 0 & 0 & 0 & 0 & 0 & 0 & 0 & 0 & 0 & 0 & 0 & 0 & 0 & 0 \\
0 & 0 & 0 & 0 & 0 & 0 & 1 & 0 & 0 & 0 & 0 & 0 & 0 & 0 & 0 & 0 \\
0 & 0 & 0 & 0 & 0 & 0 & 0 & 1 & 0 & 0 & 0 & 0 & 0 & 0 & 0 & 0 \\
0 & 0 & 0 & 0 & 1 & 0 & 0 & 0 & 0 & 0 & 0 & 0 & 0 & 0 & 0 & 0 \\
0 & 0 & 0 & 0 & 0 & 1 & 0 & 0 & 0 & 0 & 0 & 0 & 0 & 0 & 0 & 0 \\
0 & 0 & 1 & 0 & 0 & 0 & 0 & 0 & 0 & 0 & 0 & 0 & 0 & 0 & 0 & 0 \\
0 & 0 & 0 & 1 & 0 & 0 & 0 & 0 & 0 & 0 & 0 & 0 & 0 & 0 & 0 & 0 \\
0 & 0 & 0 & 0 & 0 & 0 & 0 & 0 & 1 & 0 & 0 & 0 & 0 & 0 & 0 & 0 \\
0 & 0 & 0 & 0 & 0 & 0 & 0 & 0 & 0 & 0 & 1 & 0 & 0 & 0 & 0 & 0 \\
0 & 0 & 0 & 0 & 0 & 0 & 0 & 0 & 0 & 0 & 0 & 0 & 0 & 0 & 1 & 0 \\
0 & 0 & 0 & 0 & 0 & 0 & 0 & 0 & 0 & 0 & 0 & 0 & 0 & 0 & 0 & 1 \\
0 & 0 & 0 & 0 & 0 & 0 & 0 & 0 & 0 & 0 & 0 & 0 & 0 & 1 & 0 & 0 \\
0 & 0 & 0 & 0 & 0 & 0 & 0 & 0 & 0 & 0 & 0 & 0 & 0 & 0 & 1 & 0 \\
0 & 0 & 0 & 0 & 0 & 0 & 0 & 0 & 0 & 0 & 1 & 0 & 0 & 0 & 0 & 0 \\
0 & 0 & 0 & 0 & 0 & 0 & 0 & 0 & 0 & 0 & 0 & 1 & 0 & 0 & 0 & 0
\end{bmatrix}$$

The circuit permutation matrix is given by $M = M_3 \times M_2 \times M_1 \times M_0$:

$$M = \begin{bmatrix}
1 & 0 & 0 & 0 & 0 & 0 & 0 & 0 & 0 & 0 & 0 & 0 & 0 & 0 & 0 & 0 \\
0 & 0 & 0 & 0 & 0 & 0 & 0 & 0 & 0 & 0 & 0 & 1 & 0 & 0 & 0 & 0 \\
0 & 0 & 0 & 0 & 0 & 0 & 0 & 0 & 0 & 0 & 0 & 0 & 0 & 0 & 1 & 0 \\
0 & 0 & 0 & 0 & 0 & 0 & 0 & 0 & 0 & 0 & 0 & 0 & 0 & 1 & 0 & 0 \\
0 & 0 & 0 & 0 & 1 & 0 & 0 & 0 & 0 & 0 & 0 & 0 & 0 & 0 & 0 & 0 \\
0 & 0 & 0 & 0 & 0 & 0 & 0 & 0 & 0 & 0 & 0 & 0 & 0 & 0 & 0 & 1 \\
0 & 0 & 1 & 0 & 0 & 0 & 0 & 0 & 0 & 0 & 0 & 0 & 0 & 0 & 0 & 0 \\
0 & 1 & 0 & 0 & 0 & 0 & 0 & 0 & 0 & 0 & 0 & 0 & 0 & 0 & 0 & 0 \\
0 & 0 & 0 & 0 & 0 & 0 & 0 & 0 & 1 & 0 & 0 & 0 & 0 & 0 & 0 & 0 \\
0 & 0 & 0 & 1 & 0 & 0 & 0 & 0 & 0 & 0 & 0 & 0 & 0 & 0 & 0 & 0 \\
0 & 0 & 0 & 0 & 0 & 0 & 1 & 0 & 0 & 0 & 0 & 0 & 0 & 0 & 0 & 0 \\
0 & 0 & 0 & 0 & 0 & 1 & 0 & 0 & 0 & 0 & 0 & 0 & 0 & 0 & 0 & 0 \\
0 & 0 & 0 & 0 & 0 & 0 & 0 & 0 & 0 & 0 & 0 & 0 & 1 & 0 & 0 & 0 \\
0 & 0 & 0 & 0 & 0 & 0 & 0 & 1 & 0 & 0 & 0 & 0 & 0 & 0 & 0 & 0 \\
0 & 0 & 0 & 0 & 0 & 0 & 0 & 0 & 0 & 0 & 1 & 0 & 0 & 0 & 0 & 0 \\
0 & 0 & 0 & 0 & 0 & 0 & 0 & 0 & 0 & 1 & 0 & 0 & 0 & 0 & 0 & 0
\end{bmatrix} \qquad (4.6)$$

To clarify the interpretation of the permutation matrix for a circuit, $M_{i,j} = 1$ means that input pattern j is mapped into output pattern i. For example in Eq. 4.6, $M_{1,11} = 1$ means that input pattern 11 (1011) is mapped into output pattern 1 (0001) which is readily verified from the circuit. This orientation of the permutation matrix means one can multiply M and a column vector with a 1 in the current input state (all other entries 0) and the result will be a column vector with a 1 in the resulting output state (all other entries 0).

Note that the same situation holds for multiple-valued circuits. No example is shown simply because the matrices quickly become even more unwieldy than the above small binary example.

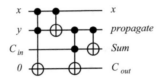

FIGURE 4.5: Reversible binary full adder.

4.4   QUANTUM GATES AND CIRCUITS

Quantum computation and quantum information have received a great deal of attention in recent years and there are a number of books on those subjects including [3,16,21,29,58,80,107]. Since our interest here concerns representing and manipulating the matrices for quantum gates and circuits, we present only a brief introduction sufficient to motivate the DD development in the next chapter. In particular, we concentrate on the mathematical aspects and not on the physical details of quantum circuits and quantum information.

The *bit* is the fundamental unit of information in conventional binary circuits. It has two possible states: 0 and 1. In quantum logic, the basic unit of information is the *quantum bit*, normally referred to as a *qubit*.

While a bit has two distinct states, a qubit state is a linear composition (superposition) of states given by:

$$|\psi\rangle = \alpha|0\rangle + \beta|1\rangle \qquad (4.7)$$

where α and β are complex numbers such that $|\alpha|^2 + |\beta|^2 = 1$. The state of a qubit is thus a vector of length 1. The notation '$|\rangle$'is known as *Dirac 'ket' notation* and is common in quantum mechanics.

The states $|0\rangle$ ($\alpha = 1, \beta = 0$) and $|1\rangle$ ($\alpha = 0, \beta = 1$) correspond to 0 and 1 for a bit. We note for completeness that it is not possible to read the state of a qubit. Rather the state is observed as 0 with probability $|\alpha|^2$ or 1 with probability $|\beta|^2$.

Quantum circuits must generally manipulate multiple qubits. The state of a pair is the following straightforward extension of Eq. 4.7:

$$|\psi\rangle| = \alpha_{00}|00\rangle + \alpha_{01}|01\rangle + \alpha_{10}|10\rangle + \alpha_{11}|11\rangle \qquad (4.8)$$

where $\Sigma_{v\in\{00,01,10,11\}}|\alpha_v|^2 = 1$. In general, the state of a system of multiple qubits is a unit vector in a Hilbert space.

The above introduced the idea of a qubit based on the two states $|0\rangle$ and $|1\rangle$. In theory, it is possible to formulate multilevel quantum digits, e.g., *qutrits* based on three states. It remains to be seen what will be possible in terms of physical realization, so we here consider the qubit case, which is the primary focus in the literature. Note that even in this case, quantum computation is considered multiple-valued, since while the inputs and outputs to a circuit may be viewed as 0 and 1, the qubit values the circuit will generally involve a discrete set with more than two values.

Any transformation of a quantum mechanical system can be specified by a unitary matrix and consequently this is the case for gates used to compose quantum circuits. Recall that a matrix M is unitary if, and only if, $M^{-1} = M^+$ (or equivalently $M^+M = I$) where M^+ denotes the conjugate transpose of M. Note that the conjugate transpose is denoted $M^\dagger$ or M^H in some works.

Permutation matrices are unitary matrices with the entries all 0 or 1. Hence, the reversible gates and circuits discussed above are in fact a special case of quantum gates and circuits.

Table 4.4 lists the unitary matrices for a selection of single qubit gates. V is called a *square root of NOT* since two V gates in a row perform the NOT operation as can be seen by squaring the matrix for V. The same property holds for the V^+ gate.

Single qubit gates are extended to create controlled gates in the same manner as shown above for binary and multiple-valued reversible gates. In particular, Algorithm 4.3 is directly applicable. The difference is that in step 1 the appropriate matrix from Table 4.8 is assigned as M_0.

An important identity for quantum gates is shown in Fig. 4.6. Here U and W are unitary operators (matrices) where $U = W^2$. For example, the matrix X for the Pauli-X (NOT) gate and the matrix V for the first $\sqrt{NOT}$ gate are such that $X = V^2$. This means a 2-control Toffoli gate can be implemented with single control gates by setting W to V in Fig. 4.6 in which case $U = X$.

This can be verified by matrix multiplication. The individual gate matrices from left to right are

$$M_0 = \begin{bmatrix} 1 & 0 & 0 & 0 & 0 & 0 & 0 & 0 \\ 0 & 1 & 0 & 0 & 0 & 0 & 0 & 0 \\ 0 & 0 & \frac{1+i}{2} & \frac{1-i}{2} & 0 & 0 & 0 & 0 \\ 0 & 0 & \frac{1-i}{2} & \frac{1+i}{2} & 0 & 0 & 0 & 0 \\ 0 & 0 & 0 & 0 & 1 & 0 & 0 & 0 \\ 0 & 0 & 0 & 0 & 0 & 1 & 0 & 0 \\ 0 & 0 & 0 & 0 & 0 & 0 & \frac{1+i}{2} & \frac{1-i}{2} \\ 0 & 0 & 0 & 0 & 0 & 0 & \frac{1-i}{2} & \frac{1+i}{2} \end{bmatrix}$$

TABLE 4.8: Matrices for selected single qubit gates

NAME	SYMBOL	MATRIX
Hadamard	H	$\frac{1}{\sqrt{2}}\begin{bmatrix} 1 & 1 \\ 1 & -1 \end{bmatrix}$
Pauli-X	X	$\begin{bmatrix} 0 & 1 \\ 1 & 0 \end{bmatrix}$
Pauli-Y	Y	$\begin{bmatrix} 0 & -i \\ i & 0 \end{bmatrix}$
Pauli-Z	Z	$\begin{bmatrix} 1 & 0 \\ 0 & -1 \end{bmatrix}$
Phase	S	$\begin{bmatrix} 1 & 0 \\ 0 & i \end{bmatrix}$
$\pi/8$	T	$\begin{bmatrix} 1 & 0 \\ 0 & e^{i\pi/4} \end{bmatrix}$
x-Rotation	$R_x(\theta)$	$\begin{bmatrix} \cos\frac{\theta}{2} & -i\sin\frac{\theta}{2} \\ -i\sin\frac{\theta}{2} & \cos\frac{\theta}{2} \end{bmatrix}$
y-Rotation	$R_y\theta$	$\begin{bmatrix} \cos\frac{\theta}{2} & -\sin\frac{\theta}{2} \\ \sin\frac{\theta}{2} & \cos\frac{\theta}{2} \end{bmatrix}$
z-Rotation	$R_z\theta$	$\begin{bmatrix} e^{-i\theta/2} & 0 \\ 0 & e^{i\theta/2} \end{bmatrix}$
$\sqrt{\text{NOT}}$ (1)	V	$\frac{1}{2}\begin{bmatrix} 1+i & 1-i \\ 1-i & 1+i \end{bmatrix}$
$\sqrt{\text{NOT}}$ (2)	V^+	$\frac{1}{2}\begin{bmatrix} 1-i & 1+i \\ 1+i & 1-i \end{bmatrix}$

$$M_1 = \begin{bmatrix} 1 & 0 & 0 & 0 & 0 & 0 & 0 & 0 \\ 0 & 1 & 0 & 0 & 0 & 0 & 0 & 0 \\ 0 & 0 & 1 & 0 & 0 & 0 & 0 & 0 \\ 0 & 0 & 0 & 1 & 0 & 0 & 0 & 0 \\ 0 & 0 & 0 & 0 & 0 & 0 & 1 & 0 \\ 0 & 0 & 0 & 0 & 0 & 0 & 0 & 1 \\ 0 & 0 & 0 & 0 & 1 & 0 & 0 & 0 \\ 0 & 0 & 0 & 0 & 0 & 1 & 0 & 0 \end{bmatrix}$$

$$M_2 = \begin{bmatrix} 1 & 0 & 0 & 0 & 0 & 0 & 0 & 0 \\ 0 & 1 & 0 & 0 & 0 & 0 & 0 & 0 \\ 0 & 0 & \frac{1-i}{2} & \frac{1+i}{2} & 0 & 0 & 0 & 0 \\ 0 & 0 & \frac{1+i}{2} & \frac{1-i}{2} & 0 & 0 & 0 & 0 \\ 0 & 0 & 0 & 0 & 1 & 0 & 0 & 0 \\ 0 & 0 & 0 & 0 & 0 & 1 & 0 & 0 \\ 0 & 0 & 0 & 0 & 0 & 0 & \frac{1-i}{2} & \frac{1+i}{2} \\ 0 & 0 & 0 & 0 & 0 & 0 & \frac{1+i}{2} & \frac{1-i}{2} \end{bmatrix}$$

$$M_3 = M_1$$

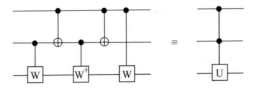

FIGURE 4.6: Circuit equivalence for $U = W^2$.

$$M_4 = \begin{bmatrix} 1 & 0 & 0 & 0 & 0 & 0 & 0 & 0 \\ 0 & 1 & 0 & 0 & 0 & 0 & 0 & 0 \\ 0 & 0 & \frac{1-i}{2} & \frac{1+i}{2} & 0 & 0 & 0 & 0 \\ 0 & 0 & \frac{1+i}{2} & \frac{1-i}{2} & 0 & 0 & 0 & 0 \\ 0 & 0 & 0 & 0 & 1 & 0 & 0 & 0 \\ 0 & 0 & 0 & 0 & 0 & 1 & 0 & 0 \\ 0 & 0 & 0 & 0 & 0 & 0 & \frac{1-i}{2} & \frac{1+i}{2} \\ 0 & 0 & 0 & 0 & 0 & 0 & \frac{1+i}{2} & \frac{1-i}{2} \end{bmatrix}$$

and the product of those matrices $M = M_4 \times M_3 \times M_2 \times M_1 \times M_0$ yields the expected matrix:

$$M = \begin{bmatrix} 1 & 0 & 0 & 0 & 0 & 0 & 0 & 0 \\ 0 & 1 & 0 & 0 & 0 & 0 & 0 & 0 \\ 0 & 0 & 1 & 0 & 0 & 0 & 0 & 0 \\ 0 & 0 & 0 & 1 & 0 & 0 & 0 & 0 \\ 0 & 0 & 0 & 0 & 1 & 0 & 0 & 0 \\ 0 & 0 & 0 & 0 & 0 & 1 & 0 & 0 \\ 0 & 0 & 0 & 0 & 0 & 0 & 0 & 1 \\ 0 & 0 & 0 & 0 & 0 & 0 & 1 & 0 \end{bmatrix}$$

Example 4.8. Figure 4.7 shows a compact circuit for the 5-bit Grover oracle function $4mod5$ which was presented in [63]. This function leaves the first four wires unchanged and inverts the last wire if, and only if, the first four wires represent an integer divisible by 5. Depending on whether the values

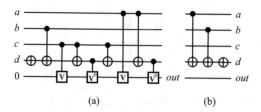

(a)   (b)

FIGURE 4.7: Circuit for Grover oracle $4 mod 5$.

of the four-input bits have to be recovered when the computation is completed, the circuit takes the form shown in Fig. 4.7(a) or that circuit followed by the input recovery gates shown in Fig. 4.7(b). The reader is welcome to verify the circuit by matrix multiplication. A more efficient approach is presented in Section 5.7.3.

Example 4.9. As a final example, consider the circuit in Fig. 4.8, which involves the following two rotation gate matrices:

$$
R_y[\pi/4](x_0) =
\begin{bmatrix}
\cos\left(\dfrac{\pi}{8}\right) & -\sin\left(\dfrac{\pi}{8}\right) & 0 & 0 & 0 & 0 & 0 & 0 \\[2mm]
\sin\left(\dfrac{\pi}{8}\right) & \cos\left(\dfrac{\pi}{8}\right) & 0 & 0 & 0 & 0 & 0 & 0 \\[2mm]
0 & 0 & \cos\left(\dfrac{\pi}{8}\right) & -\sin\left(\dfrac{\pi}{8}\right) & 0 & 0 & 0 & 0 \\[2mm]
0 & 0 & \sin\left(\dfrac{\pi}{8}\right) & \cos\left(\dfrac{\pi}{8}\right) & 0 & 0 & 0 & 0 \\[2mm]
0 & 0 & 0 & 0 & \cos\left(\dfrac{\pi}{8}\right) & -\sin\left(\dfrac{\pi}{8}\right) & 0 & 0 \\[2mm]
0 & 0 & 0 & 0 & \sin\left(\dfrac{\pi}{8}\right) & \cos\left(\dfrac{\pi}{8}\right) & 0 & 0 \\[2mm]
0 & 0 & 0 & 0 & 0 & 0 & \cos\left(\dfrac{\pi}{8}\right) & -\sin\left(\dfrac{\pi}{8}\right) \\[2mm]
0 & 0 & 0 & 0 & 0 & 0 & \sin\left(\dfrac{\pi}{8}\right) & \cos\left(\dfrac{\pi}{8}\right)
\end{bmatrix}
$$

$$R_y[-\pi/4](x_0) = \begin{bmatrix} \cos\left(\frac{\pi}{8}\right) & \sin\left(\frac{\pi}{8}\right) & 0 & 0 & 0 & 0 & 0 & 0 \\ -\sin\left(\frac{\pi}{8}\right) & \cos\left(\frac{\pi}{8}\right) & 0 & 0 & 0 & 0 & 0 & 0 \\ 0 & 0 & \cos\left(\frac{\pi}{8}\right) & \sin\left(\frac{\pi}{8}\right) & 0 & 0 & 0 & 0 \\ 0 & 0 & -\sin\left(\frac{\pi}{8}\right) & \cos\left(\frac{\pi}{8}\right) & 0 & 0 & 0 & 0 \\ 0 & 0 & 0 & 0 & \cos\left(\frac{\pi}{8}\right) & \sin\left(\frac{\pi}{8}\right) & 0 & 0 \\ 0 & 0 & 0 & 0 & -\sin\left(\frac{\pi}{8}\right) & \cos\left(\frac{\pi}{8}\right) & 0 & 0 \\ 0 & 0 & 0 & 0 & 0 & 0 & \cos\left(\frac{\pi}{8}\right) & \sin\left(\frac{\pi}{8}\right) \\ 0 & 0 & 0 & 0 & 0 & 0 & -\sin\left(\frac{\pi}{8}\right) & \cos\left(\frac{\pi}{8}\right) \end{bmatrix}$$

and three Toffoli gate matrices that are constructed as illustrated in the previous section. Multiplying the matrices appropriately yields:

$$M = \begin{bmatrix} 1 & 0 & 0 & 0 & 0 & 0 & 0 & 0 \\ 0 & 1 & 0 & 0 & 0 & 0 & 0 & 0 \\ 0 & 0 & 1 & 0 & 0 & 0 & 0 & 0 \\ 0 & 0 & 0 & 1 & 0 & 0 & 0 & 0 \\ 0 & 0 & 0 & 0 & 1 & 0 & 0 & 0 \\ 0 & 0 & 0 & 0 & 0 & -1 & 0 & 0 \\ 0 & 0 & 0 & 0 & 0 & 0 & 0 & 1 \\ 0 & 0 & 0 & 0 & 0 & 0 & 1 & 0 \end{bmatrix}$$

Note that the resulting matrix is equal to the matrix for the Toffoli gate $T(x_2, x_1; x_0)$ except for the single -1 entry. This is a relative phase factor which is sometimes encountered in physical implementations of quantum gates [80]. Phase is discussed further on the discussion of verifying circuit equivalence in Section 5.7.3.

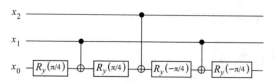

FIGURE 4.8: Quantum implementation of a Toffoli gate up to phase shift.

We conclude this chapter by considering what it means for a circuit to be reversible. We have seen that the transformation matrix for the cascade of gates $G_0 G_1 \cdots G_t$ is given by

$$M = M_t \times \cdots \times M_1 \times M_0$$

where M_i is the unitary transformation matrix for gate G_i. It is a property of the conjugate transpose of matrices that $(A \times B)^+ = B^+ \times A^+$. It is thus easily shown that

$$M^+ = M_0^+ \times M_1^+ \times \cdots \times M_t^+ \tag{4.9}$$

The importance of Eq. 4.9 is that it shows that given a cascade $G_0 G_1 \cdots G_t$ realizing a particular transformation, the inverse transformation is realized by $G_t^+ \cdots G_1^+ G_0^+$ where G_i^+ is a gate whose transformation matrix is the conjugate transpose of the transformation matrix for G_i.

From the procedure for constructing gate matrices and the target operation matrices in Table 4.7, we can see that the gate matrices for Toffoli and Fredkin gates are symmetric permutation matrices. Hence, for any circuit composed of those gates, applying the gates themselves in reverse order realizes the inverse function. However, for the MVL case, C1 and C2 must be interchanged for $p = 3$ (they are the conjugate transposes of each other). Likewise, C1 and C3 must be interchanged for $p = 4$. C2 for $p = 4$ is its own conjugate transpose.

For the quantum case, gates must be substituted by gates realizing the conjugate transpose transformation matrix, unless the gate has a matrix which is its own conjugate transpose, e.g., the Pauli-X and Hadamard gates (see Table 4.8).

The conclusion is that any reversible or quantum circuit can be transformed to realize the inverse function by reversing the order of the gates and substituting the *conjugate transpose gate* for gates where $M_i \neq M_i^+$.

CHAPTER 5

Quantum Multiple-Valued Decision Diagrams

Decision diagrams (DDs) have been previously used for storing and manipulating matrices [23,34]. Recently, Viamontes, Markov, and Hayes [117–119] presented the QuIDDPro package for handling the matrices encountered for quantum circuits. QuIDDPro employs Colorado University decision diagram (CUDD), a robust and highly efficient binary decision diagram (BDD) package developed by Somenzi [106]. QuIDDPro uses algebraic DDs [7] as implemented in CUDD, with some extensions. A major advantage of this approach is that it makes use of the extensive optimization in CUDD.

We present here an alternative approach which, since it is applicable to multiple-valued as well as binary circuits, we call *quantum multiple-valued decision diagrams* (QMDD). The goal for QMDD is to effectively represent and manipulate the complex-valued matrices encountered for quantum circuits and not complex-valued matrices in general. In particular, the regular structure of the matrices and the frequent occurrences of identity and zero submatrices are used to great advantage. As noted above, the permutation matrices required for binary or multiple-valued reversible circuits are included as a special case.

5.1 THE QMDD STRUCTURE

A $p^n \times p^n$ matrix M can be partitioned into p^2 submatrices, each of dimension $p^{n-1} \times p^{n-1}$ as shown below:

$$M = \begin{bmatrix} M_0 & M_1 & \cdots & M_{p-1} \\ M_p & M_{p+1} & \cdots & M_{2p-2} \\ \vdots & \vdots & \ddots & \vdots \\ M_{p^2-p} & M_{p^2-p+1} & \cdots & M_{p^2-1} \end{bmatrix} \tag{5.1}$$

Note that the indexing of the submatrices is from $0, \ldots, p^2 - 1$. This partitioning can be represented in a DD structure by a single vertex labeled by a selection variable with p^2 outgoing edges labeled $0, \ldots, p^2 - 1$, each pointing to a structure representing the appropriate submatrix.

The above partitioning can be performed on each submatrix and can in fact be performed repeatedly until the resulting submatrices are single elements. With no sharing of common structures and removal of redundant vertices, the result would be a decision tree. Sharing common structures and removing redundant vertices results in a DD structure, which we define formally in Definition 5.1. Note that this definition introduces the use of edge weights (complex numbers) and a particular vertex normalization rule, both of which are explained below.

Definition 5.1. A **quantum multiple-valued decision diagram (QMDD)** is a directed acyclic graph with the following properties:

1. There is a single **terminal vertex** with associated value 1. The terminal vertex has no outgoing edges.

2. There is some number of **nonterminal vertices** each labeled by a p^2-valued selection variable. Each nonterminal vertex has p^2 outgoing edges designated $e_0, e_1, \ldots, e_{p^2-1}$.

3. One vertex is the **start vertex** and has a single incoming edge that itself has no source vertex.

4. Every edge in the QMDD (including the one leading to the start vertex) has an associated complex-valued **weight**.

5. The selection variables are **ordered**, assume with no loss of generality $x_0 \prec x_1 \prec \cdots \prec x_{n-1}$, and the QMDD satisfies the following two rules:

   - Each selection variable appears at most once on each path from the start vertex to the terminal vertex.

   - An edge from a nonterminal vertex labeled x_i points to a nonterminal vertex labeled $x_j, j < i$ or to the terminal vertex. Hence, x_0 is closest to the terminal vertex and x_k for the highest k present labels the start vertex.

6. No nonterminal vertex is **redundant**, *i.e.*, no nonterminal vertex has its p^2 outgoing edges all with the same weights and pointing to a common vertex.

7. Each nonterminal vertex is **normalized** such that the largest weight on any outgoing edge is 1

8. Nonterminal vertices are **unique**, *i.e.*, no two nonterminal vertices labeled by the same x_i can have the same set of outgoing edges (destinations and weights).

Note that the start vertex is a nonterminal vertex except in the case where the QMDD has no nonterminal vertices in which case the QMDD consists of an edge pointing to the terminal vertex which is also the start vertex. This extreme case corresponds to a constant valued matrix.

Evaluating a QMDD for a particular assignment to the selection variables is equivalent to finding the value of the matrix element identified by the selection variable assignment. The evaluation involves following the path from the start vertex to the terminal vertex (including the edge leading to the start vertex) determined by the assignment. In particular, from a vertex labeled x_i follow edge e_j, where j is the value assigned to x_i. Note that some variables may not appear on the designated path. The value associated with the selection variable assignment is the product of the weights on the edges in the path.

The use of edge weights has two important effects. As a result, a QMDD has only one nonterminal vertex (with value 1) rather than a nonterminal for every unique value in the matrix. More critically, two or more submatrices that differ only by a multiplicative constant are represented by a single QMDD substructure. Efficient methods for handling the edge weights, which are complex numbers, are discussed below.

As in all DD structures, normalization is required to ensure each matrix has a unique QMDD representation. Here the normalization is quite straightforward. When a vertex is to be created to represent a matrix, the weights on the outgoing edges are examined, the largest weight is assigned to the edge pointing to the newly created vertex, and the outgoing edge weights are all divided by the largest value. As a result the largest weight on an outgoing edge is 1, and there must be at least one edge with weight 1 from each vertex. A minor complication is the fact that there is no natural ordering of the complex numbers and hence, no intrinsic definition of which of the set of values is the largest. We use the following simple pragmatic definition that works well for this purpose.

Definition 5.2.   The complex number a is **greater** than the complex number b, if $|a| > |b|$, or $|a| = |b|$, and $\angle a < \angle b$ where $\angle x$ denotes the angle associated with the complex number x expressed in polar coordinates.

Theorem 5.1.   A $p^n \times p^n$ complex valued matrix M has a unique (up to variable reordering or relabeling) QMDD representation.

Proof.   The proof is by induction on n.

When $n = 0$, the matrix is a single element, the QMDD representation of which is an edge pointing to the nonterminal vertex (with value 1) with weight equal to the matrix element value. This representation is obviously unique.

Assume the selection variable ordering $x_0 \prec x_1 \prec \cdots \prec x_{n-1}$ from the terminal vertex to the start vertex and suppose the theorem holds for all matrices up to dimension $p^{n-1} \times p^{n-1}$. To construct the QMDD, for the matrix M in Eq. 5.1, we must consider two cases:

1. If all the M_i's, $0 \le i \le p^2 - 1$ are the same, the QMDD for M is in fact the QMDD for M_0, which is unique, since the structure of M is independent of the top level selection variable and a vertex involving that variable would be redundant. Note that repeated application of this case results in the fact that any constant matrix, regardless of dimension, is represented

by an appropriately weighted single edge pointing to the terminal vertex. This includes zero matrices, a result that is quite important for achieving computational efficiency.

2. If all the M_i's, $0 \leq i \leq p^2 - 1$ are not the same, create QMDD with a start vertex labeled x_{n-1} with edges pointing to the QMDD representations of the submatrices M_i, $0 \leq i \leq p^2 - 1$. By the inductive hypothesis, the QMDD for the submatrices are unique. The vertex just created must be normalized as described in the definition of QMDD. That process is clearly unique as it involves labeling the edge pointing to the new vertex by the largest outgoing edge weight (as given by Definition 5.2), which is unique, followed by dividing all outgoing edge weights by that largest value. Since the created vertex points to unique QMDD for the submatrices and the normalization process is unique, the constructed QMDD is unique. ■

Example 5.1. Figure 5.1 shows the QMDD representation for the gate $V(x_2; x_1)$ (x_2 is the control and x_1 is the target) for a circuit with $n = 3$ lines. x_0 is an unconnected line. The 0,1,2,3 edges from each vertex are from left to right. The weights are shown next to each edge. The corresponding matrix is given in Eq. 5.2.

$$M = \begin{bmatrix} 1 & 0 & 0 & 0 & 0 & 0 & 0 & 0 \\ 0 & 1 & 0 & 0 & 0 & 0 & 0 & 0 \\ 0 & 0 & 1 & 0 & 0 & 0 & 0 & 0 \\ 0 & 0 & 0 & 1 & 0 & 0 & 0 & 0 \\ 0 & 0 & 0 & 0 & \frac{1+i}{2} & 0 & \frac{1-i}{2} & 0 \\ 0 & 0 & 0 & 0 & 0 & \frac{1+i}{2} & 0 & \frac{1-i}{2} \\ 0 & 0 & 0 & 0 & \frac{1-i}{2} & 0 & \frac{1+i}{2} & 0 \\ 0 & 0 & 0 & 0 & 0 & \frac{1-i}{2} & 0 & \frac{1+i}{2} \end{bmatrix} \qquad (5.2)$$

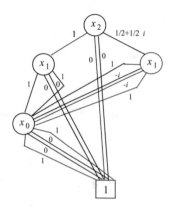

FIGURE 5.1: QMDD for $V(x_2; x_1)$ with $n = 3$ (see Eq. 5.2).

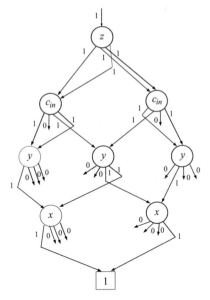

FIGURE 5.2: QMDD for adder circuit in Fig. 4.5.

Example 5.2.   As a larger example, the QMDD in Fig. 5.2 corresponds to the binary reversible adder circuit in Fig. 4.5. Selection variable z denotes the input that is constant 0 in the circuit. The edge weights are all 0 or 1 since the matrix for this circuit is a permutation matrix. the numbers on the edges are the weights. Edges from each vertex are ordered 0,1,2,3 from left to right. Note that 'dangling' edges with weight 0 point to the terminal vertex. This is not shown to simplify the diagram.

Each vertex in a QMDD is the start vertex of a subgraph that represents a submatrix within the matrix represented by the QMDD. Much of the compactness of the QMDD representation, as is the case for all DDs and as can be seen in Fig. 5.2, comes from the fact that common subgraphs (submatrices) are represented only once.

5.2   QMDD OPERATIONS

A variety of matrix operations can be efficiently performed directly on QMDD. The procedures are based on Bryant's *apply operation* [13]. The following notation is used:

- p is the radix of the logic system being considered, so every nonterminal vertex has p^2 outgoing edges;
- $w(e)$ denotes the weight on edge e;
- $v(e)$ denotes the vertex which edge e points to;

- $x(e)$ denotes the variable that labels the vertex which edge e points to ($x(e)$ returns a result that compares as earlier in the order than any variable if e points to the terminal vertex);
- $E_i(e)$ denotes the ith edge out of the vertex that e points to; and
- $term(e)$ denotes a Boolean test that is true if edge e points to the terminal vertex. The variables adhere to the same order in all QMDD and one variable comes *higher* in that order than another if the first is closer to the start vertex.

Algorithm 5.1 *(QMDD Matrix Addition).*

Let e_0 and e_1 be two edges pointing to two QMDD (matrices) to be added. The procedure is recursive and involves the following steps:

1. If $term(e_1)$, swap e_0 and e_1.
2. If $w(e_0) = 0$, the result is e_1.
3. If $v(e_0) = v(e_1)$, the result is an edge pointing to $v(e_0)$ with weight $w(e_0) + w(e_1)$.
4. Set s to identify which of $x(e_0)$ and $x(e_1)$ is higher in the variable order.
5. For $i = 0, 1, \ldots, p^2 - 1$
   (a) If $term(e_0)$ or $x(e_0) \neq s$, set $q_0 = e_0$; else set q_0 to point to $v(E_i(e_0))$ and $w(q_0) = w(e_0) \times w(E_i(e_0))$.
   (b) If $term(e_1)$ or $x(e_1) \neq s$, set $q_1 = e_1$; else set q_1 to point to $v(E_i(e_1))$ and $w(q_1) = w(e_1) \times w(E_i(e_1))$.
   (c) Recursively invoke this procedure to add q_0 and q_1 giving z_i.
6. The result is an edge pointing to a vertex labeled s with outgoing edges z_i, $i = 0, 1, \ldots, p^2 - 1$. This vertex and the edge pointing to it are normalized as described above.

Algorithm 5.2 *(QMDD Matrix Multiplication).*

Let e_0 and e_1 be two edges pointing to two QMDD (matrices) to be multiplied. The procedure is recursive and involves the following steps:

1. If $term(e_1)$, swap e_0 and e_1.
2. If $w(e_0) = 0$, the result is e_0.
3. If $term(e_0)$, the result is an edge pointing to $v(e_1)$ with weight $w(e_0) \times w(e_1)$.
4. Set s to identify which of $x(e_0)$ and $x(e_1)$ is higher in the variable order.

   For $i = 0, p, 2p, \ldots, (p-1)p$
      For $j = 0, 1, \ldots, p - 1$
         Set z_{i+j} to be an edge with weight 0
         pointing to the terminal vertex.

For $k = 0, 1, \ldots, r - 1$

    (i) If $term(e_0)$ or $x(e_0) \neq s$, set q_0 to e_0

        else set q_0 to point to $v(E_{i+k}(e_0))$

        with weight $w(e_0) \times w(E_{i+k}(e_0))$.

    (ii) If $term(e_1)$ or $x(e_1) \neq s$, set q_1 to e_1

        else set q_1 to point to $v(E_{j+p \times k}(e_1))$

        with weight $w(e_0) \times w(E_{j+p \times k}(e_1))$.

    (iii) Recursively invoke this procedure to

        multiply the QMDD pointed to by q_0 and q_1

        and then use the procedure above to add

        the result to the QMDD pointed to by z_{i+j}.

        The result of the addition replaces z_{i+j}.

5. The result is an edge pointing to a vertex labeled s with outgoing edges z_i, $i = 0, 1, \ldots, p^2 -$ 1. This vertex and the edge pointing to it are normalized as described above.

The Kronecker product of two matrices A, with dimension $q \times r$, and B, with dimension $s \times t$, is defined as

$$A \otimes B = \begin{bmatrix} a_{11}B & a_{12}B & \cdots & a_{1r}B \\ a_{21}B & a_{22}B & \cdots & a_{24}B \\ \vdots & \vdots & \ddots & \vdots \\ a_{q1}B & a_{42}B & \cdots & a_{qr}B \end{bmatrix}$$

$A \otimes B$ has dimension $qs \times rt$. Note that the Kronecker product is not commutative.

Algorithm 5.3 *(QMDD Kronecker Product)*.

Let e_0 and e_1 be two edges pointing to two QMDD (matrices) for which we want to compute the Kronecker product. For the application considered here, the selection variables for B precede the selection variables for A in the variable order. This greatly reduces the complexity of the algorithm for computing the Kronecker product of two QMDD. The procedure is recursive and involves the following steps:

1. If $w(e_0) = 0$, the result is e_0.

2. If $term(e_0)$, the result is an edge pointing to $v(e_1)$ with weight $w(e_0) \times w(e_1)$.

3. For $i = 0, 1, \ldots, p^2 - 1$, recursively invoke this procedure to find the Kronecker product of $E_i(e_0)$ and e_1 setting z_i to the result.

4. The result is an edge e pointing to a vertex labeled $x(e_0)$ with outgoing edges z_i, $i = 0, 1, \ldots, p^2 - 1$. This vertex and the edge pointing to it are normalized as described above. After normalization, $w(e)$ is set to $w(e) \times w(e_0)$.

The transpose and conjugate transpose of a matrix are very similar operations as can be seen from the following algorithm.

Algorithm 5.4 *(QMDD Transpose and Conjugate Transpose)*.
Let e be an edge pointing to a QMDD (matrix) for which we wish to compute the transpose or conjugate transpose. The procedure is recursive and involves the following steps:

1. If $term(e)$, the result is e for the transpose (use e with the weight set to the complex conjugate of $w(e)$ for the conjugate transpose.)

2. For $i = 0, 1, \ldots, p - 1$
   For $j = i, i + 1, \ldots, p - 1$
     Set $z_{i+j \times p}$ to point to the transpose (or conjugate transpose)
       of $E_{j+i \times p}$ by recursively calling this procedure
     If $i \neq j$, set $z_{j+i \times p}$ to point to the transpose (or conjugate
       transpose) for $E_{i+j \times p}$ by recursively calling this procedure.

3. The result is an edge a pointing to a vertex labeled $x(e)$ with outgoing edges z_i, $i = 0, 1, \ldots p^2 - 1$. This vertex and the edge pointing to it are normalized as described above. After normalization, $w(a)$ is set to $w(a) \times w(e)$ for the transpose (use complex conjugate of $(w(a) \times w(e))$ for the conjugate transpose).

5.3   QMDD GATE MATRIX CONSTRUCTION

Given the above QMDD procedures for matrix operations, Algorithm 4.1 for constructing the transformation matrix for a reversible gate can be efficiently implemented using QMDD. Equation 4.4, which applies for unconnected lines, is implemented using the Kronecker product of an identity matrix with the matrix constructed thus far. As noted earlier, the procedure is applicable for quantum gates using the appropriate single qubit matrix from Table 4.8 for the initial matrix.

Algorithm 4.1 assumes the targets that involve the lowest order selection variables, *i.e.*, variables closest to the terminal vertex. To allow for gates with targets in arbitrary positions, we must be able to reorder the variables in a QMDD. An approach based on the reordering method for MDD is used. This is described below.

It is possible, for single target gates, to build the QMDD directly with the target in the correct position [76]. This procedure is three to four times more efficient but is rather complex. A detailed discussion is beyond the scope of this presentation. The interested reader should consult the reference.

5.4   ROW AND COLUMN VECTORS

It is useful to be able to represent and manipulate row and column vectors. Both are readily integrated into the QMDD approach with only minor changes. Note that the vectors of interest have dimension p^n.

A row vector R of dimension p^n can be partitioned into p subvectors each of dimension p^{n-1} as follows:

$$R = [R_0, R_1, \ldots, R_{p-1}]$$

if we let ϕ denote an empty matrix of appropriate dimension, we can consider R as a pseudo-matrix giving

$$R = \begin{bmatrix} R_0 & R_1 & \cdots & R_{p-1} \\ \phi & \phi & \cdots & \phi \\ \vdots & \vdots & \ddots & \vdots \\ \phi & \phi & \cdots & \phi \end{bmatrix}$$

As above, the R_i can be recursively decomposed down to single elements. The ϕ entries are not decomposed. This idea maps readily onto the QMDD structure by introducing *null edges* for the ϕ submatrices. Column vectors can be represented in an analogous fashion.

This approach of representing vectors as pseudo-matrices by employing null edges may appear wasteful, particularly for larger values of p. In fact, it is quite a reasonable approach and allows vector–vector and vector–matrix operations to be handled by the matrix operation routines with only minor additions to account for null edges. For example, when adding two QMDD and one is a null edge, the result is the other QMDD. When multiplying or taking the Kronecker product of two QMDD, the result is a null edge if either of the two QMDD is a null edge. This saves writing specialized routines for handling each type of vector–vector or vector–matrix operation.

The transpose and conjugate transpose operate as described above with the proviso that the result of applying the operations on a QMDD that is a null edge is a null edge. It is readily seen that given this modification, the procedures above transpose a row vector to a column vector and *vice versa*.

5.5   VARIABLE REORDERING FOR QMDD

The basic sifting approach (see Algorithm 3.2) can be applied to QMDD. What is required is a procedure for adjacent variable interchange.

Algorithm 3.3 is applicable with two modifications. First the number of edges from a vertex is p^2 rather than p. Second, and more significant, step 3 specifies that for MDD, edge operations must be normalized. For QMDD, it is the complex-valued weights on the edges that must be appropriately adjusted.

A QMDD adjacent variable interchange for the binary ($p = 2$) case is depicted in Fig. 5.3. Diagram (a) shows the structure before the interchange. The edges from each vertex are ordered 0,1,2,3 from left to right. The α and β are the complex-valued edge weights. The vertices are normalized so that there is at least one weight on an edge from each vertex that equals 1, and those that do not equal 1 must have a value less than 1. The m_i at the bottom of the diagram represent the QMDD for the 16 submatrices that follow this structure. The m_i are not necessarily distinct. Note that the diagram can be interpreted to cover the case of a *skipped* y vertex in which case the corresponding β values and the m_i descendants are equal. The path following the jth edge from the top vertex (labeled x) and the kth edge from the appropriate y vertex leads to m_i, where $i = 4j + k$.

Figure 5.3(b) shows the structure after the interchange of x and y. As described in Algorithm 3.3, the top vertex is relabeled y but is physically the same vertex as the top vertex in diagram (a). This is so that edges pointing to this vertex remain valid after the interchange. The vertices labeled x in diagram (b) may be new, or may already exist.

Note the reordering of the m_i resulting from the interchange of x and y is such that the path following the jth edge from the top vertex (labeled y) and the kth edge from the appropriate x vertex leads to m_i, where $i = 4k + j$. To complete the definition of the variable interchange, we must show how the γ and δ edge weights in diagram (b) can be computed from the α and β edge weights in

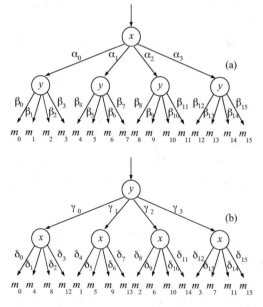

FIGURE 5.3: QMDD adjacent variable interchange.

diagram (a) such that the diagram in (b) is properly normalized and represents the same matrix as does the diagram in (a).

To begin, note that since the diagram in (a) is normalized (see Definition 5.1), the maximum value of the α and β edge weights is 1 and there must be at least one edge from each vertex in diagram (a) with edge weight 1. Set δ_i, where $i = 4j + k$ to $\alpha_k \times \beta_j$. It follows that the maximum value for these products is 1 and there must be at least one product that is 1.

Next, compute the value of each γ_i by normalizing the corresponding x vertex. This is done by dividing the four δ weights by the largest of the four and setting γ_i to that value. Clearly, the largest value assigned to any γ_i is 1 and at least one γ_i must be assigned the value 1. Hence, the structure in (b) satisfies the QMDD normalization rules. Finally, it is clear from the above construction that the m_i selected for a particular assignment to x and y is the same in each diagram, and the corresponding α, β and γ, δ pairs in the two diagrams for that assignment have the same product. Hence, the structures in the two diagrams represent the same matrix.

The above procedure can be extended to QMDD with $p > 2$. The notation gets rather complex so the detail is omitted here. It basically requires extending the above to p^2 edges for each vertex.

5.6 QMDD IMPLEMENTATION

Due to the similarity of the structures, a QMDD implementation naturally uses many of the implementation strategies described above for MDD. For example, the *unique table*, *reference counting* and *garbage collection* can be used as described for MDD with only minor modifications. The *computed table* can also be implemented as described for MDD for matrix addition, multiplication, and Kronecker product operations. It is readily extended to handle the transpose and conjugate transpose operations by taking into account the fact those operations have a single operand. As described above, the basic *sifting* approach can be applied to QMDD with the appropriate variable interchange procedure being used.

5.6.1 Identity Matrices

The use of a computed table provides significant computational advantage by avoiding duplicate computations. To further reduce computation time for matrix multiplication, the Kronecker product, and transpose operations, a flag is associated with each vertex to identify whether the QMDD structure it heads represents an identity matrix. The terminal vertex, which has value 1, represents the 1×1 identity matrix. Each nonterminal vertex implements the partition in Eq. 5.1 and represents an identity matrix if, and only if, the block matrices on the diagonal are identity matrices involving one less variable and all off diagonal entries are 0 matrices. These conditions are readily checked by examining the edges from a vertex, and the vertices those edges point to, when the vertex is entered into the unique table. Note that the check only goes to depth 1 because it is known whether each

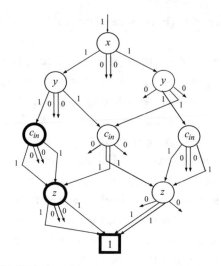

FIGURE 5.4: QMDD for adder circuit in Fig. 4.5 (reverse variable order).

immediate descendant is an identity matrix or whether it represents a 0 matrix (an edge with weight 0 pointing to the terminal vertex).

Example 5.3.   Figure 5.4 is a QMDD for the reversible binary adder in Fig. 4.5 for the reverse variable order to the one used in the construction of the matrix in Eq. 4.6 and the QMDD in Fig. 5.2. The highlighted vertices in Fig. 5.4 correspond to identity submatrices.

It is interesting to note that the two QMDD in Figs. 5.2 and 5.4 have the same number of vertices but quite different structures. Figure 5.4 has identity submatrices as shown, whereas Fig. 5.2 has no identity submatrices although the QMDD started by the leftmost vertex labeled c_{in} and the two vertices labeled x do represent diagonal matrices. This is a possible indication that a more involved metric than simple vertex count might be beneficial for variable reordering. This is an open question.

5.6.2   Edge Weights

Edge complements for BDD require only a single bit of information be added to the data structure representing an edge. Edge operators for MDD also require the addition of only a small amount of information, typically on the order of $\lceil \log_2 p \rceil$ bits. Often the required information is stored in *unused* bits in other fields so there is no increase in the memory required to represent a vertex.

The complex-valued edge weights for QMDD, require different techniques since a complex-value requires significant memory area, and the same weights typically occur many times throughout a QMDD. These facts would lead to significant redundancy and excessive memory usage if the actual weights were stored as part of the edges.

A solution, based on a technique introduced in QuIDDPro [119], is to use a simple linear list of complex values and to have each edge in the QMDD represent the weight as an index into this list rather than storing the weight itself. When a complex number is required, this list is searched and the value is inserted into the list if it is not found. The search can be sped up by arranging the list using the complex number ordering given in Definition 5.2. The approach can also be sped up by requiring that the values 0 and 1 be in positions 0 and 1 in the list, respectively. This is particularly the case for reversible circuits where 0 and 1 are the only weights required.

5.6.3   Complex Number Representation

The most straightforward method for representing complex numbers is to store the real and the imaginary parts as floating-point values. This has two disadvantages. Computations using floating-point are time-consuming, and significant round-off problems can arise particularly given the amount of computation required for larger problems.

For reversible circuits, no complex values are needed and the matrices contain only values 0 and 1. For circuits composed of *NOT(Pauli-X)*, *controlled-NOT*, *V* and V^+ gates, it is readily shown that the required complex numbers all have real and imaginary parts which are rational numbers. A rational can be accurately stored as two integers and computation over the rationals can be performed with no round-off error. Several other quantum gates can be included without moving beyond rational real and imaginary parts for the complex numbers.

However, when certain rotation gates are used, rationals are not sufficient. The transformation matrices for rotation about θ involve $\sin \frac{\theta}{2}$ and $\cos \frac{\theta}{2}$. Rotation by 0 or π is not a problem as the sin and cos values involved are 0 or 1. However, rotation by $\frac{\pi}{2}$, introduces $\sin \frac{\pi}{4}$ and $\cos \frac{\pi}{4}$ both of which equal $\frac{\sqrt{2}}{2}$, which is irrational. In this case, it transpires that the real and imaginary parts of the complex numbers involved can all be represented as quadratic irrational numbers of the form $\frac{a+b\sqrt{2}}{c}$ where a, b, and c are integers. Once again, these numbers can be stored and manipulated in the integer domain with no round-off errors. The computations are somewhat involved but are faster, and of course more accurate, than using floating-point.

The above does not handle all cases. For example, rotation by $\frac{\pi}{4}$ involves the sin and cos of $\frac{\pi}{8}$, which has no finite representation and in this case floating-point must be used.

From the above, it is clear that the complex number representation could be chosen based on knowledge of the values that will be encountered in the course of the problem. A good solution, which to date has not been implemented, would be a complex number package that adapts as values are encountered.

Two less desirable approaches that are used are (i) to represent complex numbers using the highest precision floating-point numbers available in the language being used, for example, long doubles in C, and (ii) to use a multiprecision floating-point package. Approach (i) requires appropriate tolerance operations so that round-off errors do not result in two values that should be considered the same being treated as different. Even with that, round-off can cause difficulty for problems requiring very large amounts of computation. Approach (ii) addresses the round-off issue but at considerable computational expense. More research on the representation of complex numbers is certainly needed.

5.6.4 Complex Number Computation Tables

Since operations on complex numbers are rather expensive, regardless of the representation used, and the same computations arise frequently, further computational speedup can be achieved using addition, multiplication, and division lookup tables. Subtraction is not required for QMDD. The table for each table is square and has a row and column for each complex number required. Initially each table is empty and they are populated dynamically as complex numbers are required.

Each time a complex number operation must be evaluated, the appropriate table entry is checked. If it is not empty, the computation has been performed previously and the index to the result in the complex value table is found in the appropriate operation table. If the operation table entry is empty, the required computation is performed and the index to the result is placed in the table for future reference. Commutativity holds for addition and multiplication so that only the entries above and on the diagonal are employed in those computation tables.

Typical problems for which QMDD have been applied to date require no more than 100 distinct complex values so the computation tables are not large. Note that they are indexed and store complex number indices to the complex value list and not complex numbers directly. Given this it is practical to use statically allocated arrays for these tables as well as for the complex number list. Should it later be found that a significantly larger number of complex values is required, dynamic allocation techniques would likely be more effective.

5.6.5 Unique and Computed Table Hash Functions

The hash functions used for the unique and computed tables should both take into account the edge weights on the operand edges, and the hash function for the computed table should also take into account the operation type, in order to minimize hash collisions. Designing effective hash functions can be difficult. Examples are presented here for readers interested in this level of detail.

The work reported in [74] used the following C code

```
{\tt for(i=0;i<Nedge;i++) key+=(((int)e.p->e[i].p)>>i+e.p->e[i].w);
  key=(key)&HASHMASK;}
```

for the unique table hash function where $e.p$ is a pointer to the vertex being hashed, $e.p-> e[i].p$ is the address of the vertex the ith edge points to, and $e.p-> e[i].w$ is the complex-value table index of the weight for that edge. Note the number of slots is a power of two so $key\&HASHMASK$ can be used in place of a mod operation. *HASHMASK* is the number of hash slots minus 1. The result (*key*) is a value from 0 to *HASHMASK*.

The same work used the following C code for the computed table hash function:

{\tt key=((((int)a.p+(int)b.p)>>3)+(int)a.w+(int)b.w+(int)which)&CTMASK}

where *a.p* and *b.p* are the addresses for the two operand QMDD and *a.w* and *b.w* are the corresponding weight indices. *which* identifies the type of the operation. *CTMASK* is the number of compute table slots minus 1 where again the number of slots is a power of two.

5.7   QMDD APPLICATIONS

Result reported in [74] demonstrates the effectiveness of QMDD. The experiments were performed using a QMDD package implemented in C and were run on a laptop computer with a 1.73 GHz Intel Pentium M processor and 1GB RAM running LINUX on a 256MB virtual machine under VMware 5.5. he gcc 4.0.0 C compiler with level 4 optimization to compile the QMDD package. LINUX was chosen in order to compare the results to those for QuIDDPro 3.0(beta) [119], which is available as an executable only.

5.7.1   Binary Circuits

Results for a number of binary functions from Maslov's benchmark web site [60] are reported in Table 5.1. The following information is given for each circuit:

- type which is NCT: circuit uses NOT, controlled-NOT and Toffoli gates; or NCV: circuit uses NOT, controlled-NOT, V and V^+ gates;
- number of lines in the circuit;
- number of gates in the circuit;
- number of vertices before sifting;
- time to build the QMDD in CPU milliseconds determined using the standard library time.h routines;
- number of vertices after sifting;
- time to sift the QMDD in CPU milliseconds determined using the standard library time.h routines;
- percentage vertex count reduction by sifting;
- maximum number of vertices encountered during sifting – this is an indicator of how large the QMDD might be but is not necessarily the maximum since sifting does not consider all variable orderings.

TABLE 5.1: QMDD experimental results: binary circuits

NAME	TYPE	LINES	GATES	VERTICES BEFORE SIFTING	TIME TO BUILD QMDD (MS)	VERTICES AFTER SIFTING	TIME TO SIFT QMDD (MS)	VERTEX COUNT REDUCTION BY SIFTING (%)	MAX. VERTICES DURING SIFTING	QUIDDPRO VERTICES	TIME TO BUILD BDD (MS)
5mod5	NCT	6	17	28	0	16	8	43.9	28	45	77
6symd2	NCT	10	20	247	7	170	42	31.2	478	299	117
9symd2	NCT	12	28	229	7	184	50	19.7	558	445	194
c2	NCT	35	116	150	30	136	289	9.3	504	348	1546
c2	NCV	35	305	150	220	136	241	9.3	504	348	10245
c3-17	NCT	3	6	10	0	10	4	0.0	11	21	22
c410184	NCT	14	46	39	0	33	43	15.4	86	86	274
c410184	NCV	14	74	39	0	33	52	15.4	86	86	492
cyc17-3	NCT	20	48	236	10	42	131	82.2	418	584	880
ham3	NCT	3	5	10	0	10	0	0.0	10	21	21
ham15	NCT	15	132	4522	140	2638	488	42.7	13878	7547	2051
hwb4	NCT	4	11	22	0	20	4	9.1	22	45	43
hwb4	NCV	4	21	22	0	20	8	9.1	22	45	83
hwb7	NCT	7	289	179	30	155	16	13.4	181	351	2117
hwb8	NCT	8	614	343	120	280	28	18.4	351	684	6254
hwb9	NCT	9	1541	683	640	520	40	23.9	690	1349	24566
hwb10	NCT	10	3595	1331	2920	960	59	27.9	1347	2650	98704
hwb11	NCT	11	9314	2639	14290	1730	130	34.4	2685	5223	478555
hwb12	NCT	12	18393	5167	53350	3185	265	38.4	5254	10283	1722050
rd84d1	NCT	15	28	3588	27	396	117	89.0	4415	1252	255

The results show that the effect of sifting varies significantly from example to example. A low improvement can be a result of having started from what is already a good ordering, the fact that the sifting heuristic does not visit all possible variable orderings, or that the function's QMDD representation is insensitive to variable ordering.

The results for the "hidden-weight-bit" problems hbw4–hbw12 are interesting. They show that the size of the QMDD can grow exponentially with the number of lines in the circuit. The benefit gained by sifting also increases with the number of lines.

Table 5.1 also shows the results of using QuIDDPro Version 3.0(beta) on the same computer. On average for the circuits shown, the number of vertices for the QuIDDPro representation is 2.06 times the number for the QMDD representation prior to sifting. This is as expected since a nonterminal QuIDDPro vertex has two outgoing edges while a nonterminal QMDD vertex has four outgoing edges for binary functions. What is interesting is how much the ratio can differ from 2. The largest sized DDs for the circuits shown is for cyc17_3, where the ratio is 2.47. QuIDDPro uses the highly efficient CUDD decision diagram package and also offers considerably more functionality than the QMDD implementation used in these experiments. QuIDDPro is designed for binary reversible and quantum gates and circuits.

5.7.2   Ternary Circuits

There are as yet no established benchmarks for multiple-valued reversible and quantum circuits available in the literature. This is largely because CAD tools for designing and simulating such circuits are not generally well developed. Indeed, it is hoped that QMDD will be helpful in this regard. Table 5.2 presents some ternary examples. The first is the reversible ternary adder from [71] shown in Fig. 4.2. The initial QMDD is relatively small (23 vertices) but even in this case, sifting results in notable reduction.

The S circuits are highly regular. An S circuit with n lines has $n - 1$ gates, where gate g_i is a C1 gate with target x_i and a single 1-control x_{i+1}. As expected, given this regular and quite simple structure, the QMDDs are small (the number of vertices is twice the number of lines in the circuit) and can be shown to have a very regular structure. Sifting results in no improvement, but this is a disadvantage of a heuristic, and considerable computation is required.

Each R$n - m$ circuit has n lines and m pseudo-randomly generated gates. Each gate is randomly chosen to be C1 or C2 with a randomly chosen target and a single randomly chosen control. The control is randomly chosen to be a 1 or 2-control. The improvement by sifting is, as expected, quite variable.

These examples indicate that QMDD construction and sifting are reasonably practical for quite large binary and ternary problems. It is a concern that the cost of sifting seems quite high for large ternary examples. It is likely that the implementation of sifting can be improved, but it is also the case that the matrices involved, and hence, the QMDD, are simply very complex. To put

TABLE 5.2: QMDD experimental results: ternary circuits

NAME	LINES	GATES	VERTICES BEFORE SIFTING	TIME TO BUILD QMDD (MS)	VERTICES AFTER SIFTING	TIME TO SIFT QMDD (MS)	VERTEX COUNT REDUCTION BY SIFTING (%)	MAX. VERTICE DURING SIFTING
Adder	4	16	23	0	15	4	34.8	39
S25	25	24	50	11	50	312	0.0	184
S50	50	49	100	58	100	1370	0.0	384
S75	75	74	150	168	150	3112	0.0	584
S100	100	99	200	370	200	5550	0.0	784
R5-25	5	25	132	11	124	15	6.1	185
R5-50	5	50	204	35	200	23	2.0	205
R5-75	5	75	196	58	193	19	1.5	208
R5-100	5	100	203	85	198	23	2.5	213
R10-25	10	25	1308	74	442	105	66.2	1308
R10-50	10	50	9670	687	6839	1081	29.3	12151
R10-75	10	75	41170	6776	34991	7901	15.0	45826
R10-100	10	100	51133	12361	46556	10646	9.0	52785

this in better context, note that constructing a QMDD for a p-valued, n-line circuit with m gates is equivalent to constructing $m\,p^n$ matrices and performing $m - 1$ matrix multiplications, which is a very large computational task.

5.7.3   Circuit Equivalence and Verification

QMDD can be used to quickly verify that two (or more) circuits perform the same function. All that is required is to build the QMDD for the first circuit and to then build the QMDD for the second circuit using the same variable ordering. Due to the uniqueness of the QMDD representation, the two circuits perform the same function if, and only if, they have identical QMDD and consequently the edges pointing to the top vertices for the QMDD must be identical. It is thus sufficient to check the equality of those two edges. In particular, a full traversal of the QMDD is not required.

Example 5.4.   This approach can be used to show the two adder circuits in Fig. 5.5 realize the same function. It is important to remember that the two QMDD must be built using the same variable order since the QMDD representation is only unique up to variable order.

It is critical to appreciate that this approach verifies that the two circuits realize the same reversible function. When, as in this case, a nonreversible function (the adder) has been embedded into a reversible function, that embedding must be the same for the circuits being compared. How to determine that two circuits are equivalent up to the nonreversible function for two different embeddings is a difficult and unsolved problem.

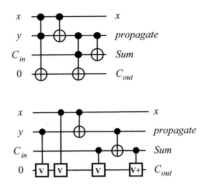

FIGURE 5.5: Two binary full adder circuits.

Example 5.5. Consider the quantum circuit in Fig. 5.6, which is a particular code error syndrome circuit []. This circuit has 13 lines and 36 gates. Note that for compactness, a single vertical line is used to show a particular control line is connected to multiple gates. The circuit in Fig. 5.7, which has 13 lines and 38 gates, is equivalent, which can be verified by comparing their QMDD representations.

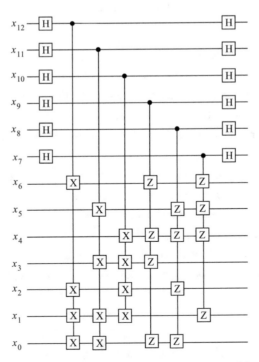

FIGURE 5.6: Code error syndrome circuit [?].

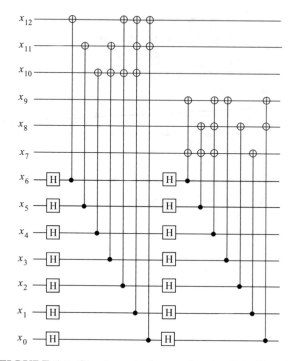

FIGURE 5.7: Circuit equivalent to the circuit in Fig. 5.6.

Using the QMDD package reported in [74], it takes about 250 ms. CPU time to build the QMDD for the circuit in Fig. 5.6 is about 100 ms. CPU time to build the QMDD for the circuit in Fig. 5.7 both from simple text files listing the gates. The difference in times is because the second circuit is mainly comprised of controlled-NOT gates. Once the two QMDD are built, confirming that they are equal takes virtually no time since, as noted above, it simply involves testing the equality of two edges.

This example is a good illustration of the power of QMDD. There are a total of 74 gates. A total of 72 matrix multiplications is required with 70 intermediate matrices constructed. Hence the problem deals with 146 matrices each of which has dimension $2^{13} \times 2^{13}$. The CPU time required is small, about 1/3 of a second. Memory usage is also not very high. The QMDD has 452 vertices and the unique table finishes with 15,633 vertices. No garbage collection is invoked. Each vertex occupies 44 bytes including the space for the pointers and complex number list indices for the outgoing edges. The complex value list has 25 entries after both diagrams are built.

Circuits are considered equivalent if they implement the same function. For quantum circuits, it is often useful to know if two circuits are equivalent up to global or relative phase shift [80] rather than strictly equivalent. This can be determined by finding M_1 and M_2, the transformation matrices for the two circuits, and computing $M_1 \times M_2^+$. The two circuits are equivalent up to phase if the

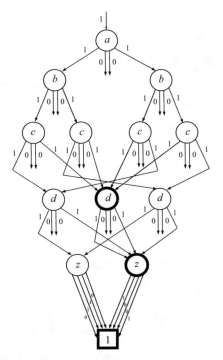

FIGURE 5.8: QMDD for Grover $4mod5$ oracle (Figure 4.7).

resulting product matrix is a diagonal matrix. Since the required operations are available, QMDD can be used to verify circuit equivalence up to phase shifts.

QMDD can also be applied to the problem of circuit verification, i.e., verifying that a circuit behaves according to a specific functional description.

Example 5.6. Consider the Grover $4mod5$ oracle circuit given in Fig. 4.7 including the input recovery gates. The QMDD is shown in Fig. 5.8. The QMDD package gives the matrix and truth table in Table 5.3 which for compactness are shown in text form as produced by the package. Note that in the package, matrices are displayed with periods rather than 0's to emphasize permutation patterns.

It can be seen from the matrix and truth table in Table 5.3 that the circuit inverts the value of z when $abcd$ is a 4-bit number evenly divisible by 5. Since z is constant 0, the circuit produces a 1 on the corresponding output when the required condition is met. The package output verifies that the input recovery gates function correctly.

Verification by means of a matrix and truth table works well for circuits with a small number of lines. For larger problems, one useful approach is to build a QMDD directly from a functional specification and to then compare that to the QMDD built from the circuit. It is relatively easy to

TABLE 5.3: Matrix and truth table for Grover 4*mod*5 oracle circuit

PERMUTATION MATRIX	ABCDZ	
.1.....................	00001	1
1......................	00000	0
..1....................	00010	2
...1...................	00011	3
....1..................	00100	4
.....1.................	00101	5
......1................	00110	6
.......1...............	00111	7
........1..............	01000	8
.........1.............	01001	9
...........1...........	01011	11
..........1............	01010	10
............1..........	01100	12
.............1.........	01101	13
..............1........	01110	14
...............1.......	01111	15
................1......	10000	16
.................1.....	10001	17
..................1....	10010	18
...................1...	10011	19
.....................1.	10101	21
....................1..	10100	20
......................1	10110	22
.......................1	10111	23
........................1	11000	24
.........................1	11001	25
..........................1	11010	26
...........................1	11011	27
............................1	11100	28
.............................1	11101	29
...............................1	11111	31
..............................1.	11110	30

build a QMDD from a truth table or cube list description, for example, since, as shown in Chapter 3, those representations are formulated using logical operations that are directly computable on QMDD. The presentation above should provide the reader sufficient implementation detail to build QMDD-based CAD software applicable to specific verification tasks.

CHAPTER 6

Summary

MVL is a topic that has many applications both in conventional computer systems design and in emerging technology areas such as quantum logic and computer design.

In Chapter 1, we surveyed several applications of MVL including usage in CAD-EDA algorithms and MVL circuit design. An underlying concept in the applications surveyed is that MVL is not limited to only nonbinary applications. The CAD-EDA applications discussed in Chapter 1 included algorithms for binary logic simulation, synthesis, testing, and verification. Although each of these algorithms targets conventional binary logic, the underlying algorithms utilize concepts of MVL internally, which often yield better results than those that would be obtained if only binary logic were utilized.

In terms of circuit design, we provided examples of the use of MVL that support both binary and MVL circuits. For the case of binary digital logic, it was pointed out that while the target circuitry utilizes binary signal values, the theory behind the design of such circuitry is based on MVL. From this point of view, such circuits are inherently based on MVL and the ultimate realization is simply an encoding of the logic values as binary strings. The choice of encoding MVL values as a set of nonbinary signal levels or as sets of binary strings is guided by characteristics of the device technology to be used as basic components in the resulting circuitry. Because MOS transistors are better suited for binary switching logic, most circuits in use currently are based on two-voltage levels; however, it is certainly the case that principles of MVL are used in their design. The specific examples in Chapter 1 that are based on MVL principles but utilize binary string encoding are addition circuits based on nonbinary redundant digit sets and PLA architectures that are based on quaternary logic. A major goal of these examples is to convince the reader that MVL concepts have important applications for the design of logic circuits and that the ultimate implementation of those circuits using binary-valued signals is merely an issue of logic value encoding.

Chapter 1 also contained a discussion of current and emerging devices that utilize multiple (more than two) voltages to encode discrete logic values. In terms of emerging devices, the resonant tunneling diode (RTD) and the single-electron transistor (SET) are discussed. Both of these devices exploit the effect of quantum tunneling and are well suited for implementing MVL circuits. Examples of conventional circuitry that use more than two voltage or current values are MOS transistor based current mode logic and memory circuits based on a special type of floating gate MOS transistor that can be used to construct memory cells that store multiple discrete voltages. These two examples illustrate the importance of MVL for emerging and modern circuitry.

The historical basis of the development of MVL was surveyed from a mathematical point of view in Chapter 2. The earliest work considered was developed by discrete mathematicians whose application interests were in the development of logic systems that were more expressive than Boolean systems for reasoning. Emphasis is placed upon the concept of functional completeness of such systems as this is a very important characteristic when such systems are used as the foundation for logic circuitry and algorithms supporting computer systems. More recent logic systems were included in Chapter 2 whose developments were motivated primarily to support logic circuit design in an efficient manner. These latter systems also included logic systems whose basic operators can be described in terms of modular arithmetic and thus also have applications in discrete communication system protocols and in information theory.

Chapter 3 considered representations of MVL systems that allow for the manipulation of operations to support efficient algorithm design. The chapter began with a discussion of truth tables arranged in a traditional fashion and then discusses truth tables arranged in the form of a map. The map arrangement allows for the concept of a cube, or conjunction of literals, to be developed and methods for determining function cubes for both the binary and MVL case are described. Next, the concept of a cube list and algorithms for the manipulation of cube lists were described. The final representation structure considered is that of the decision diagram (DD). DDs allow for the compact representation of a function as a directed graph structure and also allow for the formulation of efficient algorithms for their manipulation. Both binary- and multiple-valued DDs were considered and algorithms for the creation of these data structures are described as well as support structures such as the unique table and the computed table. An important topic with regard to DDs is the minimization of the structures through variable reordering. A discussion was presented regarding the implementation of the "sifting" method for diagram size reduction and corresponding garbage collection mechanisms concluded the chapter.

Chapter 4 was devoted to the topic of reversible and quantum logic circuits. This topic is becoming more prevalent in the literature due to the fact that reversible logic circuits can approach the lower limit of zero-power dissipation based on thermodynamic information entropy principles. Binary reversible circuits are first introduced followed by MVL versions. Next, quantum logic circuits were described. Quantum logic circuits all have the characteristic of being composed as a cascade of reversible operators. Because the basic unit of information in a binary quantum logic circuit, the qubit, and the basic unit of information in a MVL quantum circuit, the qudit, were modeled as vectors in finite dimensional Hilbert spaces, MVL and nonbinary methods were used for their description. In particular, individual quantum logic gates can be modeled using unitary transformation matrices to describe their functionality.

The focus of Chapter 5 was on representation methods for quantum MVL circuits; specifically the quantum multiple-valued decision diagram (QMDD). These DDs are fundamentally different from those considered in Chapter 3 since they have vertex decomposition rules based on the structure of the transformation matrix being represented rather than an algebraic property such as the Shannon decomposition used for binary decision diagrams. Because quantum circuits can be modeled as matrices of size $p^k \times p^k$ where p is the radix value of the logic system and k is the number of qudits, it was very important to use a compact representation of such circuits for the purpose of design tasks such as simulation, synthesis, and verification. Algorithms were described for the manipulation of quantum circuits represented as QMDDs including variable reordering through sifting. Chapter 5 concluded with a presentation of experimental results that provide the reader with information regarding the effectiveness of the data structure in terms of the required memory and the speed of representing a quantum circuit as a QMDD.

In summary, it is the intention of the authors to provide an introduction to the area of MVL, particularly as it applies to computer system design issues both from a circuit design and an algorithmic point of view. A description of the usefulness of MVL in both binary and higher logic systems is a principal goal and this material should equip the reader with the necessary background to understand and utilize more advanced topics in the area of MVL, including reversible and quantum logic circuits.

BIBLIOGRAPHY

[1] A. Agrawal and N. K. Jha. Reversible logic synthesis. In *Proc. Design, Automation and Test in Europe*, pp. 1384–1385, 2004.

[2] S. B. Akers. Binary decision diagrams. *IEEE Trans. Comput.*, vol. 27, pp. 509–516, 1978, doi:10.1109/TC.1978.1675141.

[3] A. N. Al-Rabadi. *Reversible Logic Synthesis: From Fundamentals to Quantum Computing.* Springer-Verlag, 2003.

[4] C. M. Allen and D. D. Givone. A minimization technique for multiple-valued logic systems. *IEEE Trans. Comput.*, vol. 17, pp. 182–184, 1968, doi:10.1109/TC.1968.227407.

[5] W. C. Athas and L. J. Svensson. Reversible logic issues in adiabatic CMOS. In *Proc. PhysComp '94*, 1994, pp. 111–118.

[6] A. Avizienis. Signed-digit number representations for fast parallel arithmetic. *IRE Trans. Electronic Comput.*, vol. 10, pp. 389–400, 1961.

[7] R. I. Bahar, E. A. Frohm, C. M. Gaona, G. D. Hachtel, E. Macii, A. Pardo, and F. Somenzi. Algebric decision diagrams and their applications. *Formal Methods Syst. Design: An Int. J.*, vol. 10, no. 2-3, pp. 171–206, 1997.

[8] D. Beatty, R. E. Bryant, and C. -J. H. Seger. Formal hardware verification by symbolic ternary trajectory evaluation. In *Proc. Design Automat. Conf.*, 1991, pp. 397–402, doi:10.1023/A:1008699807402.

[9] B. A. Bernstein. Modular representations of finite algebras. In *Proc. 7th Int. Congress Mathematicians, Univ. Toronto Press, 1928, vol. I*, 1924, pp. 207–216.

[10] D. A. Bochvar. Ob odnom tréhznačhom isčislénii i égo priménénii k analizu paradoksov kalssičéskogo rassirénnogo funkcional'nogo isčislenija (tr. On a 3-valued logical calculus and its application to the analysis of contradictions). *Matématičsekij sbornik*, vol. 4, pp. 287–308, 1939.

[11] K. S. Brace, R. L. Rudell, and R. E. Bryant. Efficient implementation of a BDD package. In *Proc. Design Automat. Conf.*, 1990, pp. 40–45.

[12] R. E. Bryant. MOSSIM: A switch-level simulator for MOS LSI. In *Proc. Design Automat. Conf.*, 1981, pp. 786–790.

[13] R. E. Bryant. Graph-based algorithms for Boolean function manipulation. *IEEE Trans. Comput.*, vol. 35, no. 8, pp. 677–691, 1986, doi:10.1109/TC.1986.1676819.

[14] R. E. Bryant. Extraction of gate level models from transistor circuits by four valued symbolic analysis. *IEEE Trans. CAD*, 10(2), pp. 350–353, 1991.

[15]   R. E. Bryant, D. Beatty, K. Brace, K. Cho, and T. Sheffler. COSMOS: A compiled simulator for MOS circuits. In *Proc. Design Automat. Conf.*, 1987, pp. 9–16.

[16]   I. Burda. *Introduction to Quantum Computation*. Universal Publishers, 2005.

[17]   J. T. Butler (editor). *Multiple-Valued Logic in VLSI Design*. IEEE Computer Society Press, 1991.

[18]   F. Capassa, S. Sen, F. Beltram, L. M. Lunardi, A. S. Vengurlekar, P. R. Smith, N. J. Shah, R. J. Malik, and A. Y. Cho. Resonant-tunneling transistors, circuits with reduced complexity and multiple-valued logic. *IEEE Trans. Electron Devices*, vol. 36, no. 10, pp. 2065–2082, 1989, doi:10.1109/16.40888.

[19]   R. D. Chamberlain and M. A. Franklin. Collecting data about logic simulation. *IEEE Trans. CAD*, vol. 5, no. 7, pp. 405–411, 1986.

[20]   L. L. Chang, L. Esaki, and R. Tsu. Resonant tunneling in semiconductor double barriers. *Appl. Phys. Letters*, vol. 24, pp. 593–595, 1974, doi:10.1063/1.1655067.

[21]   G. Chen. *Mathematics of Quantum Computation and Quantum Technology*. Taylor & Francis, 2007.

[22]   E. M. Clarke, O. Grumberg, and D. A. Peled. *Model Checking*. The MIT Press, 1999.

[23]   E. M. Clarke, K. Mcmillan, X. Zhao, M. Fujita, and J. Yang. Spectral transforms for large boolean functions with applications to technology mapping. *Formal Methods Syst. Design: An Int. J.*, vol. 10, no. 2-3, pp. 137–148, 1997, doi:10.1023/A:1008695706493.

[24]   M. Cohn. *Switching Function Canonical Forms over Integer Fields*. Ph.D. Thesis, Harvard University, 1960.

[25]   A. De Vos, B. Raa, and L. Storme. Generating the group of reversible logic gates. *J. Physics A: Math. Gen.*, vol. 35, pp. 7063–7078, 2002, doi:10.1088/0305-4470/35/33/307.

[26]   K. Degawa, T. Aoki, T. Higuchi, H. Inokawa, and Y. Takahashi. A single-electron transistor logic gate family and its application – Part I: Basic components for binary, multiple-valued and mixed-mode logic. In *Proc. Int. Symp. Multiple-valued Logic*, 2004.

[27]   E. Dubrova and J. C. Muzio. Generalized Reed-Muller canonical form for a multiple-valued algebra. *Multiple Valued Logic – An Int. J.*, vol. 1, pp. 65–84, 1996.

[28]   G. W. Dueck, D. Maslov, and D. M. Miller. Transformation-based synthesis of networks of Toffoli/Fredkin gates. In *Proc. IEEE Canadian Conf. Electrical and Computer Engineering*, 2003.

[29]   K. Engesser. *Handbook of Quantum Logic and Quantum Structures: Quantum Structure*. Elsevier Science and Technology, 2007.

[30]   G. Epstein. *Multiple-Valued Logic Design: An Introduction*. IOP Publishing Ltd, 1993.

[31]   H. A. Castro *et al.* A 125MHz burst mode 0.18μm 128 Mbit 2 bits per cell flash memory. In *Proc. VLSI Symp. Technology Circuits*, 2002.

[32]   D. Etiemble and M. Israel. Comparison of binary and multivalued ICs according to VLSI criteria. *IEEE Comput.*, pp. 28–42, 1988.

[33]   E. Fredkin and T. Toffoli. Conservative logic. *Int. J. Theor. Phys.*, vol. 21, pp. 219–253, 1982, doi:10.1007/BF01857727.

[34]   M. Fujita, P. C. McGeer, and J. C. -Y. Yang. Multi-terminal binary decision diagrams: an efficient data structure for matrix representation. *Formal Methods Syst. Design: An Int. J.*, vol. 10, no. 2-3, pp. 149–169, 1997, doi:10.1023/A:1008647823331.

[35]   M. R. Garey and D. S. Johnson. *Computers and Intractability – A Guide to NP-Completeness.* Freemann, 1979.

[36]   M. Gunes, M. A. Thornton, F. Kocan, and S. A. Szygenda. A survey and comparison of digital logic simulators. In *Proc. Midwest Symp. Circuits Syst.*, 2005.

[37]   T. Hanyu, Y. Yabe, and M. Kameyama. Multiple-valued programmable logic array based on a resonant tunneling diode model. *IEICE Trans. Electron.*, vol. E76-C, pp. 1126–1132, 1993.

[38]   Y. Harata, Y. Nakamura, H. Nagese, M. Takigawa, and N. Takagi. A high-speed multiplier using a redundant binary adder tree. *IEEE J. Solid-State Circuits*, vol. 22, pp. 28–34, 1987, doi:10.1109/JSSC.1987.1052667.

[39]   M. A. Heap and M. R. Mercer. Least upper bound on OBDD sizes. *IEEE Trans. Comput.*, vol. 43, pp. 764–767, 1994, doi:10.1109/12.286311.

[40]   C. W. Hemming and S. A. Szygenda. Modular requirements for digital logic simulation at a predefined level. In *Proc. ACM Annu. Conf.*, pp. 380–389, 1972.

[41]   W. N. N. Hung, X. Song, G. Yang, J. Yang, and M. Perkowski. Quantum logic synthesis by symbolic reachability analysis. In *Proc. Design Automat. Conf.*, 2004, pp. 838–84, doi:10.1109/TC.1984.1676392.

[42]   S. L. Hurst. Multiple-valued logic – its status and future. *IEEE Trans. Comput.*, vol. 33, pp. 1160–1179, 1984.

[43]   S. L. Hurst. *The Logical Processing of Digital Signals.* Crane-Russack, 1978.

[44]   The Simucad Inc. Silos user's manual 2002.1. Technical report, Simucad Inc., 2002.

[45]   H. Inokawa and Y. Takahashi. Experimental and simulation studies of single-electron transistor based multiple-valued logic. In *Proc. Int. Symp. Multiple-valued Logic*, 2003.

[46]   M. Kameyama, S. Kawahito, and T. Higuchi. A multiplier chip with multiple-valued bidirectional current-mode logic circuits. *IEEE Comput.*, pp. 43–56, 1988, doi:10.1109/2.50.

[47]   M. Karnaugh. The map method for synthesis of combinational logic circuits. *Trans. of A.I.E.E.*, vol. 72, no. pt. 1, pp. 593–599, 1953.

[48]   P. Kerntopf. Synthesis of multipurpose reversible logic gates. In *Proc. EUROMICRO Symp. Digital Systems Design*, 2002, pp. 259–266.

[49] P. Kerntopf. A new heuristic algorithm for reversible logic synthesis. In *Proc. Design Automat. Conf.*, 2004, pp. 834–837.

[50] S. C. Kleene. On a notation for ordinal numbers. *The Journal of Symbolic Logic*, vol. 3, pp. 150–155, 1938, doi:10.2307/2267778.

[51] S. Kuninobu, T. Nishiyama, T. Edamatsu, T. Taniguchi, and N. Takagi. Design of high speed mos multiplier and divider using redundant binary representation. *IEEE Trans. Comput.*, vol. 36, pp. 80–85, 1987.

[52] R. Landauer. Irreversibility and heat generation in the computing process. *IBM J. Res. Dev.*, vol. 5, pp. 183–191, 1961.

[53] C. Y. Lee. Representation of switching circuits by binary decision diagrams. *Bell Syst. Tech. J.*, vol. 38, pp. 985–999, 1959.

[54] J. -S. Lee, Y. Chung, J. Kim, and S. Lee. A practical method of constructing quantum combinatorial logic circuits. Technical report, arXiv:quant-ph/9911053v1, 1999.

[55] K. K. Likharev. Single-electron devices and their applications. *Proc. IEEE*, vol. 87, no. 4, pp. 606–632, 1999, doi:10.1109/5.752518.

[56] J. Lukasiewicz. O logice tròjwartościowej (tr. on three-valued logic). *Ruch Filozoficzny*, vol. 5, pp. 169–171, 1920.

[57] A. B. Marcovitz. *Introduction to Logic Design*. McGraw-Hill, 2nd ed., 2005.

[58] D. C. Marinescu and G. M. Marinescu. *Approaching Quantum Computing*. Pearson Education, 2004.

[59] N. M. Martin. The Sheffer functions of 3-valued logic. *J. Symbolic Logic*, vol. 19, no. 1, pp. 45–51, 1954, doi:10.2307/2267650.

[60] D. Maslov. Reversible logic synthesis benchmarks page. http://www.cs.uvic.ca/~dmaslov/, 2005.

[61] D. Maslov and G. W. Dueck. Asymptotically optimal regular synthesis of reversible networks. In *Proc. Int. Workshop Logic Synthesis*, 2003, pp. 226–231.

[62] D. Maslov, G. W. Dueck, and D. M. Miller. Simplification of Toffoli networks via templates. In *Symp. Integrated Circuits and System Design*, pp. 53–58, 2003.

[63] D. Maslov, C. Young, D. M. Miller, and G. W. Dueck. Quantum circuit simplification using templates. In *Proc. Design, Automation and Test in Europe*, pp. 1208–1213, 2005.

[64] P. Mazumder, M. Bhattacharya, J. Sun, and G. Haddad. Digital circuit applications of resonant tunneling diodes. *Proc. IEEE*, vol. 86, no. 4, pp. 664–685, 1998, doi:10.1109/5.663544.

[65] E. J. McCluskey. Minimization of Boolean functions. *Bell Syst. Tech. J.*, vol. 35, 1956.

[66] E. J. McCluskey. *Introduction to the Theory of Switching Circuits*. McGraw-Hill, 1965.

[67] K. L. McMillan. *Symbolic Model Checking*. Kluwer Academic Publisher, 1993.

[68] C. Meinel, F. Somenzi, and T. Theobald. Linear sifting of decision diagrams. In *Proc. Design Automat. Conf.*, 1997, pp. 202–207.

[69] D. M. Miller and R. Drechsler. Implementing a multiple-valued decision diagram package. In *Proc. Int. Symp. Multiple-valued Logic*, 1998.

[70] D. M. Miller and R. Drechsler. Augmented sifting of multiple-valued decision diagram. In *Proc. Int. Symp. Multiple-valued Logic*, 2003, pp. 275–282.

[71] D. M. Miller, G. Dueck, and D. Maslov. A synthesis method for MVL reversible logic. In *Proc. Int. Symp. Multiple-valued Logic*, 2004, pp. 74–80.

[72] D. M. Miller and G. W. Dueck. On the size of multiple-valued decision diagrams. In *Proc. Int. Symp. Multiple-valued Logic*, 2003, pp. 235–240.

[73] D. M. Miller and G. W. Dueck. Spectral techniques for reversible logic synthesis. In *6th Int. Symp. Representations and Methodology of Future Computing Technologies*, 2003, pp. 56–62.

[74] D. M. Miller, D. Y. Feinstein, and M. A. Thornton. Variable reordering and sifting for qmdd. In *Proc. Int. Symp. Multiple-valued Logic (CD-Rom)*, 2007.

[75] D. M. Miller, D. Maslov, and G. W. Dueck. A transformation based algorithm for reversible logic synthesis. In *Proc. Design Automat. Conf.*, 2003, pp. 318–323.

[76] D. M. Miller and M. A. Thornton. QMDD: A decision diagram structure for reversible and quantum circuits. In *Proc. Int. Symp. Multiple-valued Logic (CD-Rom)*, 2006.

[77] S. Minato, N. Ishiura, and S. Yajima. Shared binary decision diagrams with attributed edges for efficient Boolean function manipulation. In *Proc. Design Automat. Conf.*, 1990, pp. 52–57.

[78] D. E. Muller. Application of Boolean algebra to switching circuit design and error detection. *IRE Trans.*, vol. 1, pp. 6–12, 1954.

[79] J. C. Muzio and T. C. Wesselkamper. *Multiple-Valued Switching Theory*. Adam Hilger, Bristol and Boston, 1986.

[80] M. Nielsen and I. Chuang. *Quantum Computation and Quantum Information*. Cambridge Univ. Press, 2000.

[81] C. S. Peirce. A boolean algebra with one constant. In C. Hartshorne and P. Weiss, editors, *Collected Papers of Charles Sanders Peirce*, vol. 4, pp. 12–20. Harvard University Press, 1980.

[82] M. Perkowski, L. Jozwiak, P. Kerntopf, A. Mishchenko, A. Al-Rabadi, A. Coppola, A. Buller, X. Song, M. M. H. A. Khan, S. Yanushkevich, V. Shmerko, and M. Chrzanowska-Jeske. A general decomposition for reversible logic. In *5th Int. Reed–Muller Workshop*, 2001, pp. 119–138.

[83] E. L. Post. Introduction to a general theory of elementary propositions. *Amer. J. Math.*, vol. 43, pp. 163–185, 1921, doi:10.2307/2370324.

[84] D. K. Pradhan. A multivalued switching algebra based on finite fields. In *Proc. Int. Symp. Multiple-valued Logic*, 1974, pp. 95–111.

[85] W. Quine. A way to simplify truth functions. *Amer. Math. tmonthly*, vol. 62, pp. 627–631, 1955, doi:10.2307/2307285.

[86] I. S. Reed. A class of multiple-error-correcting codes and their decoding scheme. *IRE Trans. Information Theory*, vol. 3, pp. 6–12, 1954.

[87] N. Rescher. *Many-valued Logic*. McGraw-Hill Education, 1969.

[88] D. C. Rine(editor). *Computer Science and Multiple-Valued Logic Theory and Applications*. North-Holland Publishing Company, 1977.

[89] J. P. Roth. Diagnosis of automata failures: A calculus and a method. *IBM J. Res. Dev.*, vol. 10, pp. 278–281, 1966.

[90] J. P. Roth. Algebraic topological methods for the synthesis of switching systems I. *Trans. American Mathematical Society*, vol. 88, no. 2, pp. 301–326, 1958, doi:10.2307/1993216.

[91] C. Rozon. On the use of VHDL as a multi-valued logic simulator. In *Proc. Int. Symp. Multiple-valued Logic*, 1996, pp. 110–115.

[92] R. Rudell. Dynamic variable ordering for ordered binary decision diagrams. In *Proc. Int. Conf. CAD*, 1993, pp. 42–47.

[93] A. Salz and M. Horowitz. IRSIM: An incremental MOSswitch-level simulator. In *Proc. Design Automat. Conf.*, 1989, pp. 173–178.

[94] T. Sasao. Multiple-valued decomposition of generalized boolean functions and the complexity of programmable logic arrays. *IEEE Trans. Comput.*, vol. 30, pp. 635–643, 1981, doi:10.1109/TC.1981.1675861.

[95] T. Sasao. Input variable assignment and output phase optimization of PLA's. *IEEE Trans. Comput.*, vol. 33, pp. 879–894, 1984, doi:10.1109/TC.1984.1676349.

[96] T. Sasao. Multiple-valued logic and optimization of programmable logic arrays. *IEEE Comput.*, pp. 71–80, 1988.

[97] T. Sasao. *Switching Theory for Logic Synthesis*. Kluwer Academic Publishers, 1999, doi:10.1109/2.52.

[98] The ModelSim SE. PSL Assertions guide. Tech. Rep., Model Technology Inc., 2002.

[99] C. J. H. Seger. VOSS – a formal hardware verification system user's guide. Tech. Rep., 93-45, Department of Computer Science, University of British Columbia, 1993.

[100] C. -J. H. Seger and R. E. Bryant. Formal verification by symbolic evaluation of partially-ordered trajectories. *Formal Methods Syst. Design: An Int. J.*, vol. 6, pp. 147–189, 1995, doi:10.1007/BF01383966.

[101] H. M. Sheffer. A set of five independent postulates for Boolean algebra with application to logical constants. *Trans. Am. Math. Soc.*, vol. 14, pp. 481–488, 1913, doi:10.2307/1988701.

[102] V. V. Shende, A. K. Prasad, I. L. Markov, and J. P. Hayes. Reversible logic circuit synthesis. In *Proc. Int. Conf. CAD*, 2002, pp. 125–132.

[103] J. Slupecki. Der volle dreiwertige aussagenkalkül (tr. the full three-valued propositional calculus). *Comptes rendus des séances de la Société des sciences et des lettres de Varsovie*, Classe III, vol. xxix, pp. 9–11, 1936.

[104] J. Slupecki and L. Borkowski. *Elements of Mathematical Logic and Set Theory*. PWN, 1967.

[105] K. C. Smith. Multiple-valued logic: A tutorial and appreciation. *IEEE Comput.*, pp. 17–27, 1988.

[106] F. Somenzi. CUDD: CU decision diagram package – release 2.4.1. http://vlsi.colorado.edu/~fabio/CUDD/cuddIntro.html, 2005.

[107] K. Svozil. *Quantum Logic*. Springer-Verlag, Berlin, 1998.

[108] S. A. Szygenda. TEGAS2 – Anatomy of a general purpose test generation and simulation system for digital logic. In *Proc. 9th Workshop on Design Automation*, 1972, pp. 116–127.

[109] S. A. Szygenda and Thompson. Digital logic simulation in a time-based, table-driven environment, parts I and II. *IEEE Comput.*, vol. 8, 1975.

[110] N. Takagi and S. Yajima. On-line error-detectable high-speed multiplier using redundant binary representation and three-rail logic. *IEEE Trans. Comput.*, vol. 36, no. 11, pp. 1310–1317, 1987.

[111] Y. Takahashi, A. Fujiwara, Y. Ono, and K. Murase. Silicon single-electron devices and their applications. In *Proc. Int. Symp. Multiple-valued Logic*, 2000.

[112] Y. Takahashi, Y. Ono, A. Fujiwara, and H. Inokawa. Silicon single-electron devices. *J. Phys. Condensed Matter*, vol. 14, no. 39, pp. R995–R1033, 2002, doi:10.1088/0953-8984/14/39/201.

[113] H. Tang and H. Lin. Multi-valued decoder based on resonant tunneling diodes in current tapping mode. In *Proc. Int. Symp. Multiple-valued Logic*, 1996, pp. 230–234.

[114] The Intel Corporation. Intel®StrataFlash™Memory Technology. Application Note AP-677, 1998.

[115] T. Toffoli. Reversible computing. Tech Memo LCS/TM-151, MIT Lab for Comp. Sci., 1980.

[116] R. Vasquez, J. Juan-Chico, M. Bellido, A. Acosta, and M. Valencia. HALOTIS: High accuracy logic timing simulator with inertial and degradation delay model. In *Proc. Design, Automation and Test in Europe*, pp. 467–471, 2001.

[117] G. F. Viamontes, I. L. Markov, and J. P. Hayes. Graph-based simulation of quantum computation in the state-vector and density-matrix representation. In *Proc. SPIE*, vol. 5436, 2004.

[118] G. F. Viamontes, I. L. Markov, and J. P. Hayes. Graph-based simulation of quantum computation in the density matrix representation. *Quantum Information & Computation*, vol. 5, no. 2, pp. 113–130, 2005.

[119] G. F. Viamontes, I. L. Markov, and J. P. Hayes. QuIDDPro: High-performance quantum circuit simulation. http://vlsicad.eecs.umich.edu/Quantum/qp/, 2005.

[120] Z. G. Vranesic, E. S. Lee, and K. C. Smith. A many-valued algebra for switching systems. *IEEE Trans. Comput.*, vol. 19, pp. 964–971, 1970, doi:10.1109/T-C.1970.222803.

[121] T. Waho, K. Chen, and M. Yamamoto. Literal gate using resonant-tunneling devices. In *Proc. Int. Symp. Multiple-valued Logic*, 1996, pp. 68–73.

[122] D. L. Webb. Generation of any *n*-valued logic by one binary operator. *Proc. Natl. Acad. Sci.*, vol. 21, pp. 252–254, 1935.

[123] S. Yang. Logic synthesis and optimization benchmarks user guide. Technical Report 1/95, Microelectronic Center of North Carolina, 1991.

[124] A. A. Zinov́ev. *Philosophical Problems of Many-Valued Logic*. D. Reidel Publishing Company, 1963, doi:10.1073/pnas.21.5.252.

Author Biography

D. Michael Miller was born in Winnipeg, Manitoba. He received a B.Sc. in Mathematics and Physics from the University of Winnipeg in 1971 and a M.Sc. and Ph.D. in Computer Science from the University of Manitoba in 1973 and 1976. From 1975 to 1980, he was a faculty member in the School of Computer Science, University of New Brunswick and moved to the University of Winnipeg in 1980 and then to the University of Manitoba in 1982.

Dr. Miller joined the Department of Computer Science at the University of Victoria in 1987 as Chair, a position held for a decade except for the Fall of 1992 when he was Acting Dean of Engineering, and July 1995 - June 1996 when on administrative leave at TIMA Laboratory/INPG, Grenoble, France. He has been Dean of the Faculty of Engineering since July 1997, except for an administrative leave from August to December 2002 in the Faculty of Computer Science at the University of New Brunswick in Fredericton.

In addition to his administrative duties at the University of Victoria, he is treasurer of the Victoria Section of APEGBC and Chair of the National Council of Deans of Engineering and Applied Sciences (NCDEAS). Dean Miller is a member of the Association for Computing Machinery (ACM) and the Institute of Electrical and Electronics Engineers (IEEE). He is the Secretary for the IEEE Computer Society Technical Committee on Multiple-Valued Logic.

His current primary research interests include the synthesis of reversible and quantum logic circuits, and decision diagrams applied to the design of binary and multiple-valued logic systems using both conventional and spectral techniques.

Mitchell A. Thornton received the BSEE degree from Oklahoma State University in 1985, the MSEE degree from the University of Texas in Arlington in 1990, and the MSCS in 1993 and Ph.D. in computer engineering in 1995 from Southern Methodist University in Dallas, Texas. His industrial experience includes full-time employment at E-Systems (now L-3 communications) in Greenville, Texas and the Cyrix Corporation in Richardson, Texas where he served in a variety of engineering positions between 1985 through 1992. From 1995 through 1999, he was a faculty member in the Department of Computer Systems Engineering at the University of Arkansas and from 1999 through 2002 in the Department of Electrical and Computer Engineering at Mississippi State University. Currently, he is a Professor of Computer Science and Engineering and, by courtesy, Electrical Engineering at Southern Methodist University. His research and teaching interests are in the general area of digital circuits and systems design with specific emphasis in EDA/CAD methods including asynchronous circuit and computer arithmetic circuit synthesis, formal verification/validation and simulation of digital systems, multiple-valued logic, and spectral techniques.

Printed in the United States
by Baker & Taylor Publisher Services